No Change? No...

No Change?
No Chance!

Jean Lambert

JON CARPENTER

To Ian and Karen

My sincere thanks to Steve and Ian Lambert for their comments, typing, nagging and support: to Jon Carpenter for asking me and for his patience: to Jan Clark and John Valentine for their input and to David Morgan of the London Ecology Centre and staff at Charter 88 and the New Economics Foundation for answering my questions.

Jean Lambert
August 1996

First published in 1997 by
Jon Carpenter Publishing
The Spendlove Centre, Charlbury, Oxfordshire OX7 3PQ
☎ 01608 811969

© Jean Lambert 1996

The right of Jean Lambert to be identified as author of this work has been asserted in accordance with the Copyright, Design and Patents Act 1988

All rights reserved. No part of this publication may be reproduced, stored in a retrieval system or transmitted in any form or by any means electronic, mechanical, photocopying or otherwise without the prior permission in writing of the publisher

ISBN 1 897766 23 8

Printed in England by J.W. Arrowsmith Ltd., Bristol

Contents

Introduction		vi
1	Why bother?	1
2	Not coping on ideas	4
3	Not delivering on poverty	8
4	Failing on community	19
5	Global insecurity	31
6	Ecological insecurity	41
7	The new framework	53
8	Tackling poverty	57
9	Caring for communities	71
10	Global security	87
11	Getting the ecology right	99
12	No change, no chance	111
Bibliography		114
Index		115

Introduction

This book is about politics — and Green politics in particular. There are many books around which will help you to 'green' your lifestyle — checking labels for energy efficiency; buying local, organic produce in season; learning to love your bike and so on. This book will explain why the current system makes it so difficult to do any of these things.

There are also many books about how to develop your spiritual and conscious self, so that you may feel more at one with the planet and its inhabitants. This book will explain why, under the current system, the spiritual is undervalued.

It is also possible to find weighty, academic tomes about how Greens fit with the new social movements, and which discuss our history, and the differences between tendencies or our development in different countries.

However, there are fewer books about how and why Greens engage with politics, in the sense that most people understand politics: Parties, policy and elections. That is what this book is about.

To some, being Green and being involved in politics is philosophically incompatible — by engaging in politics you have already sold out to the system. 'Real' Greens want to do away with all that stuff and live harmoniously with nature.

Others will say that their politics, of whatever colour, are already green (an interesting visual concept!), so there is no need to create something else. 'Come and do your politics with us,' sing these Sirens. 'We will look after you — your concerns are our concerns.'

Yet others will say to those of us engaged in Green politics that the pressure groups do these things so well that there is no need, or space, for a political party.

These criticisms are not new to many of us, but they need an answer. I want to show why we have to engage directly in changing the political framework that shapes our lives and our futures: why it is not safe or responsible to leave it to the paternalistic good-will of others and why we cannot wait to perfect ourselves as individuals before we get involved. Most of all, I want to show why the changes that Greens want are worth voting for.

·1·
Why bother?

*'This outline is intended to help people visualise
what is possible in a sustainable society, not to present a blueprint...'*
Manifesto for a Sustainable Society

I joined the Green Party (the Ecology Party as it then was) in 1977 — four years after it was founded. I was drawn in by a friend (I still talk to him!) and by the political logic of the Party's position.

It was a time when we had seen the effects of the oil crisis on our economy and our way of life. We had been forced to realise how vulnerable we were to external events largely beyond our control. The nuclear power programme was expanding (under the then Labour government) and there was a growing protest movement concerned at the long-term effects on our health and our environment, as well as the obvious links with the military use.

The influential *Limits to Growth* and *Blueprint for Survival* had been published; the latter stated the need for '... a national movement, to act at a national level, and if need be to assume political status and contest the next election. It is hoped that such an example will be emulated in other countries, thereby giving rise to an international movement...'

It seemed to me then that the logic was irrefutable. Certain resources on the earth are finite: there are limits to how far we can increase the production of basic foodstuffs and maintain a clean, accessible water supply, and these resources, divided among an increasing population, mean that there will come a point where there is not enough to maintain human life — let alone that of other species. Faced with these facts, we can either carry on as before and hope that something will turn up (another agricultural green revolution, plague, space travel to find resources or whatever) or we can make choices consciously as to how we should live within these constraints.

To make such choices requires a democratic political framework and individuals or parties to act within it. No single pressure group, as they are currently understood, is equipped or designed to take on this task. Pressure groups do not have to test the support for their ideas amongst the population as a whole: their role is to influence decision-making on certain topics, not to replace the decision-makers, and generally they do not seek to effect overall change — although as individuals, their supporters may feel that to be necessary.

As far as I am concerned, the logic of Green politics still holds and is still not fully embraced by any other political movement.

There have been some changes in political thinking since 1977 and there are individual MPs and activists in other parties who do understand and support the Green political agenda, but progress has been slow, partial and often begrudging in the UK. I have the feeling that none of the changes made is secure and really entrenched in the political philosophy or policy of either likely party of government. 'Environment' is not a word that springs readily to the lips of either Tory or Labour leaders in scripted or unscripted moments.

Those born in the year I became overtly involved in politics can now vote. Like many other young people, a significant number will not bother. In the 1992 General Election over 2.5 million of those aged 18 to 25 did not make the trip to the polling station. Many young people are not even registered to vote.

Why not? Why is it that this group — theoretically the first environmentally aware generation — is failing to participate at even the most elementary level of politics?

There are many explanations. An estimated quarter of a million people voluntarily failed to register in response to the introduction of the Community Charge or Poll Tax: some in protest, some because they did not want to pay, or a mixture of both. The willingness to give up the right to vote, I believe, says a great deal about people's perception of the value of voting as a means of effecting change. A number are homeless or travellers and therefore have no fixed address at which to register.

It is also argued that those involved in non-violent direct action movements, such as the anti-roads activists, feel that their actions have greater value and are more likely to bring about change than voting. Yet this involves comparatively few young people — a number of whom vote as well.

The most common reason given for not voting is that there is 'no point'. These people feel that the political system is not interested in them and their concerns. Looking at the current 'play-it-safe' battle for 'Middle England' you can understand their point of view.

Over the last few years, young people have seen their right to income support removed, unless they are willing to undergo what many see as spurious, cheap-labour training schemes with a very limited chance of employment at the end. Students now have to be prepared to fund themselves through college or take out loans. Accommodation is hard to come by. Society — whatever that might be — expects the young to behave as responsible, independent individuals and yet denies them many of the means to do that.

The groups least likely to vote, or to take part in any other political activity (such as signing petitions) are young men from the D and E groups (as the pollsters describe them) and particularly young black men. Indeed, they are least likely to be registered as voters. These are the groups for whom employment prospects are bleakest and for whom the chances of being arrested for involvement in crime are highest. These are people who fail to see themselves represented in our political parties or other institutions — politics is 'not for us, or about us'.

The awful truth is that this is the first generation in modern history for whom we cannot say that we believe their future will be better than ours. They would not believe us if we did.

They know about shifting employment patterns, social fragmentation, the risk of war and environmental problems.

So do many other people. Poll after poll shows a growing public disillusionment with politicians and politics, which is not only to do with the 'sleaze' factor. It runs far deeper than that. There is a sense of remoteness — of those in power not really understanding what is happening on the ground and not responding to people's needs — a feeling that politics is slipping away from people and that only those with big wallets or those who shout loudly and behave aggressively are listened to.

This is dangerous. The political vacuum will be filled by those who offer a form of order and security for some and decrease it for others — whether that be the Front National in the south of France, or organised crime as in parts of East Europe.

I would contend that people's growing lack of faith in politics has everything to do with the inability of politicians to respond positively to the four broad areas of *poverty, community, global relationships* and *ecology*.

This is basically because conventional politics carries so much baggage and belief from the past that its thinking cannot provide solutions for now and for the future. Green politics stems from the here-and-now and looks forward. It therefore looks at the world in a different way and, while demanding some very fundamental changes in our values and practices, it offers a better chance of lasting security.

· 2 ·

Not coping on ideas

A question frequently asked of Greens is: 'Where do you see yourselves on the political spectrum — right or left?' Any attempt to answer by challenging the assumptions behind the question is listened to politely and then followed by another question starting, 'Yes, but...'

The shorthand of a political spectrum of only 180° has become an obsession and a strait-jacket, typifying thinking which can only cope with absolutes, opposites and opposition: a thinking which also seems to believe that progress too is a nice, clear, straight line.

Presumably, what questioners mean about the left/right divide is to do with whose interests you put first — the collective or the individual — and the degree of state (government) intervention you would employ to ensure those interests. It's a shorthand, used to cover issues such as how you define 'equality' — whether it is ensured by people *having the same* (whether that be material possessions, income, education, democratic rights) or whether it means a *meritocracy*, where those who can rise or develop will do so.

In economic terms, the left/right spectrum is understood to indicate whether one commands and directs the economy at a state level to meet the needs of the state and its people, or whether the economy can meet people's desires more effectively if left to run free of control — 'the market' (i.e. the people as consumers) deciding what will happen.

I know that there will be those reading this who say: 'Ah, but look at the writings of x, y and z: they are socialists and they have a different view. Greens are merely their natural successors!' I also know that there are Greens who happily define themselves as socialist, presumably because we are concerned with the distribution of power and with social justice.

However, I and many other Greens are not happy to see ourselves historically 'boxed' in this way and for a variety of reasons. Perhaps because it gives too little value to other political, cultural or religious roots which Greens feel are of value to them and which help to provide ideas and solutions. Also because the terms have their roots in the nineteenth century and deal with a particular set of circumstances rooted in a heavily industrialised, western society.

Most people do not describe themselves as an historical Brand A or B socialist/liberal/conservative. Their response is likely to be 'Who?' or 'What does that mean?'

People want to know what politicians stand for, what they want to do and how they will do it. Their understanding of political labels is drawn from a variety of sources and has a general set of ideas and values attached to it. I am predominantly concerned with how the general public, not the political literati, understand politics.

There are enormous shortcomings in the simple spectrum view of things, partly because the late twentieth century has added new factors — not least the reality of global ecological damage and weapons of mass destruction.

The similarities of right and left

I have always been interested by the way in which the extremes of the spectrum resemble each other in practical terms. Regimes which would define themselves and be defined externally as politically very different can behave in identical ways — political oppression of any opposition, torture, death squads, land seizure, high military expenditure, rigged 'democracy', degradation of the environment — the list goes on. The ruling group may be there through birth or political, military or religious promotion. Their actions are justified as being in 'the best interests of the country and my people'. What they have in common is an absolute belief in their arguments, their right to govern and the danger of the alternative.

There will be those who will argue (and it is usually on cold street corners when you want to do other things!) that these absolutist regimes do not really reflect the political values they espouse — they are corrupt and controlled by outside agencies and are therefore not a true reflection of that brand of politics. To some extent this is true, but it still doesn't answer the question as to why is it that, if the political spectrum is a straight line with opposites at either end, the opposites are so similar?

Probably because the straight line ignores the realities of absolute fundamentalism, whatever its roots — be they political, religious or simply the desire for power. And it is fundamentalism that brings the opposites together. A belief that only that group knows 'the truth' and that those who question that absolute truth, or who do not show sufficient fervour in proclaiming it and obeying those with the knowledge, are the 'enemy'. It is what joins McCarthy's Committee on Un-American Activities with the Algerian FPS and Stalin's purges. It justifies genocide, slavery, xenophobia, religious conflict and the oppression and slaughter of women. It is a rule founded on fear and hatred of what is different and a blinding arrogance that allows those in control to literally force others to believe as they do. It is possibly the most dangerous driving force in history and we still see it at work today.

The left/right divide does not deal comfortably, either, with the problem of scale. A driving tenet in current politics is for larger-scale organisations — usually justified by the demand for greater efficiency. This was reflected in the main parties' consensus over the need to agree the General Agreement on Trade and Tariffs (GATT) and establish the World Trade Organisation (WTO). We can see it too in the push for a wider European Union (EU) — albeit with certain caveats about its powers and the timescale for growth from the different parties and their members.

We have seen more power accruing to central government and the civil service over the last twenty or so years: it was not a process exclusive to the Conservatives, although it has certainly proceeded apace under their government. For all Labour's talk of devolution of power, it is not yet certain how far a Labour government would reverse this process and how changes would be entrenched. The Liberal Democrats have a much clearer commitment to decentralisation of power from Westminster, but a very strong commitment to the EU and WTO. There is consensus among the main parties on the continuing existence of NATO.

Where then does the concept of subsidiarity — appropriate scale — figure in a left/right analysis? It is a critical issue for the politics of the future. The view that 'bigger is better' has lead to a loss of democratic accountability, particularly in the areas of economy and community.

Also missing in the straight-line spectrum approach is the environment and its importance in the scale of things. Left and right may give indications as to the framework for its protection, but they give little indication of the value placed upon it.

Greens would argue that left and right have much more in common with each other in their view of the earth than either has with a Green perspective. Both left and right have viewed the planet and all its resources as a 'free good' — there for the benefit of all mankind (and yes, I do mean mankind: that's another perspective they share). The resources are there for us to develop and use to 'create' wealth — material goods and the means to buy them. The general attitude has been that the planet will cope with pollution and, if things get too bad, technology will come up with something or, if the proof is undeniable, a particular process may be limited or stopped. It can take a long time and a number of deaths before the latter approach takes effect.

There is an agreement, within conventional politics, that increasing consumption is good, because it improves the standard of people's lives and provides jobs, so that people have the money to consume. To trade resources, products or services is also seen as a positive good and something to be encouraged as an integral part of a sound economy.

This consensus is, of course, now running head-on into the problems of the ecological restraints on consumption and the economy. The planet is not a 'no-cost' gift. But living in harmony with the planet is not seen as a left/right issue.

Why not? Because the consensus has been that *it is not an issue at all.* One of my earliest memories as a political activist is of being at a count for the local elections of 1978 where our candidate was approached by a winning councillor. 'It's nice to see someone taking care of the environment,' our candidate was told. The implication being, of course, that it wasn't the Council's job or the job of the speaker's party (he happened to be Labour).

The toughest job that the Green Party has faced has been to make other politicians see that the environment and its security is a political issue.

Both perspectives have acted to divorce the way we live from the planet that provides our living. Their politics is about how to run the money economy.

Many Greens would argue that this stems from a view of progress which is also predominantly linear and which sees technological development as *the* measure of progress. This form of development is to do with our mastery of nature — seeing how far we can remove ourselves from its limits and vagaries. We want to see ourselves as being in control, not being controlled.

This perspective, combined with an ever-narrowing view of economics, has come to diminish our sense of what is important and thus what politics is about.

This view gives no real status to other values and views. It is a very 'western', rationalist perspective and we should not, therefore, be surprised that it cannot respond to the demands now being made upon it. People are looking for a way forward which gives value to other definitions of progress and the demands of the planet itself.

· 3 ·
Not delivering on poverty

Within the UK we are seeing a widening gap in income between the rich and the poor. While we may have record numbers of people actually in work (chiefly because we have a growing population) and the highest percentage of women in work of any EU country, yet we also have many individuals and households dependent on state benefits to keep a roof over their heads and their children fed and clothed. The government is worried at the amount we spend on social security payments and is constantly looking at ways to 'target' benefits to the most needy — meaning finding ways to cut the bills, rather than tackling the underlying problems.

These problems include the fact that employment patterns are undergoing a radical change, that many of those in work are not paid a living wage, that there are barriers in the way of many who wish to work, and that the very assumption itself that an income is provided by paid work is under serious question.

The gap between rich and poor is not only a UK phenomenon. It is apparent in many EU countries and in the United States, where levels of malnutrition and infant mortality in some sectors of the population read like Third World statistics.

The gap is there, too, in developing countries, with absolute grinding poverty at the bottom end and conspicuous consumption at the other.

Despite all the promises and programmes of the last 50 years, we still have a world where one in five people live in absolute poverty; 23 million people are classed as refugees and 13 million live as bonded slaves, due to debt.

Here, we are seeing the government and main opposition struggling to find ways to keep the promises made to many. How, for example, can we fund and provide care for an aging population?

How can we provide and maintain housing when people are having to come to terms with their home not being an appreciating asset, indeed possibly unsaleable altogether if built by particular production methods?

Homeowners see MIRAS (Mortgage Interest Relief at Source) being cut back and the DSS no longer continuing to pay interest on mortgages after October 1995. Housing benefit is being restricted to the local average rent and is no longer payable to students. Councils are now seen as housing 'enablers' and are no longer primary providers of social housing, having been forced to sell many of their properties. Housing Associations are being pushed to fill the gap and find more private sector funding, which can push rents up.

How are we to provide for our children, particularly given the increase in single-parent households, one in five of which currently survives on under £100 a week? The Child Support Agency was set up to try to cut the cost to the Treasury of providing for children of single-parent households. Its failures are well catalogued. However, its aim has been to substitute payments without necessarily augmenting them. For many heads of such households, the majority of whom are women, this has replaced a regular and assured payment with an often spasmodic method, liable to sudden change in the event, say, of the job loss of the payer. It has not made any positive material difference to the households affected.

There are also still almost half a million households with income earners who depend on family credit to supplement their income so they can provide for their children.

If, however, you ask people why they work, they are likely to reply that they do it, amongst other reasons, to 'keep a roof over their head, provide for their kids and to give them some comfort and security in their old age.' Yet we are now in a situation where, for many people, one or more of these goals is at risk. If working and paying taxes do not provide this sense of security, what will? No wonder so many people play the National Lottery...

Yet our government acts at times as if all those on benefit are work-shy, simply needing a tighter and tougher system to get them 'on their bikes' looking for work (which is about the only time such an ecological means of transport has had major backing from a senior minister!).

Yet we know there are parts of the country where paid work is very difficult to come by. We know that the skills offered by the unemployed do not always match the skills needed by employers. We also know that much of the work on offer is part-time or low paid. Research shows that many employers discriminate against those over the age of 45, or even younger in some fields. We know that if you are of Afro-Caribbean, Asian or African background you are more likely to be unemployed or over-qualified for the work that you are doing.

Still, however, you will hear people say — 'There's plenty of work about

— just look at the number of jobs advertised in the local paper.' They never seem to do the simple arithmetic of dividing the number of jobs by the number of local people registered as unemployed, let alone looking at this in a more sophisticated manner — by subtracting the number of part-time, agency (which can often be short-term) or low paid jobs first, before doing the division. Even then, you will get nothing like the full picture.

The tip of the iceberg

The official unemployment figures, as most of us are aware, only represent a proportion of those who wish to change their paid-work situation. The figures do not reflect those wishing to work part-time, to increase their working hours or to change jobs: they do not include those who have made themselves 'voluntarily redundant' by leaving paid work, nor do they include those on the various training schemes, 'redeployed' to other categories by the government.

Those wanting to work may also find themselves in the 'poverty trap' — the situation where working may leave you worse off than not working due to the benefits you may lose once you take up paid work. It seems incredible to me that a government so committed to getting people out of what they see as the dependency culture of benefits and off the state payroll, has not acted to remedy that ludicrous situation and has been incapable of effecting an integrated system.

The situation for women is even bleaker. Recent research by the Equal Opportunities Commission shows that the majority of women are still financially disadvantaged compared to men throughout their lives. Women may be less likely to be unemployed, but they are more likely to be in part-time, low paid jobs.

The situation is further complicated by the change in 'traditional' family relationships, as more women come to head households, either as single parents (90% of whom are women) or as primary earners.

The feminisation of poverty is a serious issue. Given that women provide most of the informal work that underpins the formal economy, on a limited income their ability to do this is threatened. 80% of consumer decisions are made by women — there are serious implications for the way our economy is run if women have less money to spend.

The economy is driven by maintaining and increasing consumption, whether of goods or services. Whenever we hear politicians commenting on the state of our economy, they will talk about such things as consumer confidence, the consumer-led economy or the state of play in the High Street. They will tell us that unemployment will be cured when people start buying more and borrowing more to do it. Credit levels have become a measure of confidence and therefore a 'good thing', whereas saving your money in the bank or building society is likely to be seen as a problem.

This is insanity!

One of the factors which helped to swell the spending tidal wave of the 1980s was the easy availability of credit. People of all ages were urged to borrow, regardless of their assets. In many cases they borrowed more than could be repaid or more than the goods were worth. This was particularly so in the case of houses, where changes in the rules governing tax allowances for single, joint-mortgage holders, plus the enforced sale of council houses at knock-down, subsidised prices, together with changes in the rules governing the practices of banks and building societies, combined to fuel panic-buying of houses and exorbitant increases in house prices.

Many people found themselves with mounting debts and rising job insecurity, as firms embraced new 'leaner and meaner' management methods and lending institutions developed the bad-tempered hangovers that so often follow blind binges! These institutions, guilty of poor financial management and risk assessment, then swung the other way. Worried by the provisions they were having to make for bad debts — and the effect that was having on their profits — they began to restrict lending and to call in overdrafts. For many companies and small businesses this meant disaster: the number of business failures and concomitant job losses rose rapidly, adding to the depth of the developing recession and decreasing public confidence. It became a larger version of what had happened in dairy farming in the early 1980s, when milk quotas were reduced rapidly after many farmers had been encouraged to borrow and expand their output. Such sudden rule changes can have disastrous consequences for individuals.

So many people are left with debts and have less chance of repaying them. The Citizens Advice Bureau now finds that debt counselling is the largest single demand made upon it. We are still seeing a steady number of house repossessions and people unable to move — whether from expensive Dockland developments or from ex-council blocks. It is difficult to gauge the feelings of bitterness and failure this experience has engendered.

The growth fallacy

There are many who have accused the Greens of wanting a stagnant, no-growth economy — we don't, as I shall explain later. It was left to conventional economics to deliver that.

Yet we now hear the Chancellor talking of the need for 'sustainable growth' of about 3% as a means of getting the economy back on its feet and putting people back to work. The Opposition may criticise him for failing to deliver this, but they do not question the need for it.

What is not being spelled out to the public is that conventional economic growth will not automatically deliver a 'feel-good' factor and rescue them from an insecure future. Why not?

Because economic growth is a bald, one dimensional measure, all about 'never mind the quality — feel the width'. All it measures is the amount of activity in our Gross Domestic Product (GDP). It does not tell us what the activity is, or the reasons behind it. So, if your house is burgled and you have to pay higher insurance premiums (if you can afford them, or get insurance at all) and buy new goods to replace the old, that contributes to GDP, but you do not feel happier or more secure for spending the cash. Similarly, the costs of cleaning up after flood damage or street riots may add to GDP, but they do not do much for our quality of life.

Economic growth of itself will not necessarily meet our basic needs, either physically or emotionally.

New patterns of work

Nor will it necessarily provide us with paid work. The link between growth and jobs is broken. Changes in technology and management practice mean that staff can often be replaced by machines, whether on the factory floor, the farm or the office. We now have what is known as Automatic Unemployment. Employers are looking to get more work from fewer people — what has become known as the '$1/2 \times 2 \times 3$' formula, half the people, paid twice as much, producing three times as much. The Delors White Paper, *Growth, Competition and Employment*, published by the European Commission, made it clear: between 1970 and 1992, while the growth of the European Community in real terms was 73%, employment had grown by only 7%.

Work patterns too are changing with a growing 'casualisation' of the work force as more people work on fixed term, fixed payment contracts and put together a 'portfolio' of part-time jobs — usually this means a number of well-paid contracts but for many people it represents three or four cleaning and bar-jobs. Others work in zero-hour jobs, having no fixed hours but often entailing very long hours. Others may work from home but, contrary to the rosy picture of journalists and accountants working in a hi-tech way, homeworkers are often very poorly paid outworkers for garment or toy makers who pay their own heating, lighting and storage costs saving money for their employers — very like the cottage industries before the industrial revolution.

Commentators and economists now talk of the 'work-rich' — those who are very well-paid but who work very long hours, not always because that is what the job demands but because employers expect such a tangible demonstration of staff commitment. It may in fact be counter-productive for the company as research has shown that the quality of work is likely to deteriorate in those working over about 50 hours a week: junior doctors and lorry drivers will confirm this.

There are also the 'work-rich, pay-poor' workers: the low-waged who work enormously long hours and still struggle to make ends meet, whether making samosas for a pound an hour or selling clothes for the more generous hourly rate of £1.75. Some people's weekly earnings are too low for them to pay Social Security contributions, which has serious implications for entitlement to pensions and some other benefits. These jobs are usually low- or un-skilled and often have employers who are happy to rely on a high turnover in the workforce or on part-timers. It is easier to justify low wages to part-timers as you can tell yourself that they are not the primary earner in the household, otherwise they would be working full-time, so their earnings are just pin money. This attitude can be reinforced by the fact that four out of five part-time workers are women. However, the number of men in part-time work is now rising — will this bring a change of attitude?

The 'casual' economy is more difficult to regulate in terms of employment rights and, to some extent, in terms of taxation. It is also difficult to unionise. However, it is viewed as more *efficient* — workers can be taken on and laid off as demand requires, thus reducing the wages bill. Some would see the insecurity over job prospects as a useful way of keeping wages down and workers' minds on their jobs!

Yet there are already tensions apparent in that work philosophy. Good results often require effective team work: team building requires time and trust. Short-term contracts can make this difficult to achieve unless, of course, you are taking on a whole team. How can you ensure a company ethos with people who need have no sense of loyalty or responsibility to it?

Another tension is in the area of training. Who is responsible for training and upgrading the work-force if companies are only going to permanently employ a core group?

It is generally believed that the UK television industry now depends upon the BBC to train the people who then set up the independent companies from which the BBC and other networks purchase programmes. What happens when the BBC reduces its training role?

How can we ensure the very highly skilled training of those involved in the information technology industry — particularly the consultants and the teachers?

The changes in technical and vocational training, towards emphasising the ability *to do* at the expense of also being able *to understand* has been widely criticised. Without the technical knowledge, it is more difficult for people to adapt to changes in technology as they do not have the intellectual 'tools' to comprehend the changes independently — they have become dependent on others to show them how. This may be training but it is not education. This reductionist approach also makes it more difficult for people to appreciate the cultural context of change and to evaluate its wider effect upon both their own work patterns and on society as a whole. Craft training too, has fallen dramatically, leaving us with growing skills shortages. Many skills which will be needed in a sustainable society are almost lost.

Training for jobs has become a key political promise from the three big parties and the reasoning for this is that it is 'in the national interest'. The driving force is not that it may provide people with a greater sense of self-fulfilment, give them economic choice and enable them to earn more, but that the UK must be more internationally competitive, expand its exports and encourage inward investment. Thus, training becomes your patriotic duty!

The economic future

So, what sort of world are we to be trained for? What are all these changes leading to?

Certainly a world of increasing competition, where the UK cannot assume it will have a leading role. It is deeply ironic that having been perceived as leading the move to a deregulated, free-trade economic environment, the UK is likely to become a victim of that philosophy, espoused so enthusiastically by Mrs. Thatcher and her many backers. However, we should remember that the ideas and general direction were already in circulation before her advent. No one achieves that much power alone.

The EU was already working towards a free trade region and the GATT was working towards the liberalisation of world trade — both of them organisations controlled by governments and guided by large-scale industrial and financial institutions.

The Thatcherite push towards free trade was, in many ways, building on an historical view of Britain leading the world — not least economically.

The Industrial Revolution, with its strong export of manufactured goods, and the development of London as a financial centre, had not been built on regional and state control, but on imagination, the entrepreneurial spirit and low wages. It had also been built on cheap natural resources from our colonies and our own land.

What we have come to see is an over-riding emphasis on the international economy and the neglect of the internal local and regional economy. This is what a commitment to free trade means in practice. The reason any government supports free trade is that it assumes its country will be a winner in this competition; the British government does not talk about the costs of free trade, only its 'opportunities' and 'necessities'.

What are the costs?

To succeed demands low financial costs (wages, raw materials etc.) and high efficiency — that is added value and a return on your investment: it means the flexibility to move your work to areas which will give you more of these two factors. You need to be able to move cash rapidly, and if you are moving products, success will require efficient, cheap and often rapid transport. Success requires a reliable energy and water supply. Above all, it requires expanding markets — more people buying more of your product and/or paying higher prices for it.

So free trade demands *increasing levels of consumption*. The need to keep the costs of natural resources low means that, theoretically, the providers of those resources are in a powerful position as the foundation of economic activity. In practice, however, they are the sector most heavily controlled by external forces who want to keep things cheap.

The need for transport links usually means more roads — whether to get logs out of the Amazon or yoghurt out of France. To keep costs down there will be strenuous lobbying against taxation on fuel or water costs.

Low wages become important, so there is pressure to mechanise, keep union activity out or heavily restricted and to cut permanent staff to a minimum — hence the changes in work patterns.

The permanence of the enterprise is always in question. Should circumstance change and production become cheaper elsewhere, the business will often relocate, with a sometimes devastating effect on the local community. This can also make national economies and governments more unstable as they become more dependent on external forces.

The UK government is proud that this country has more inward investment than any other EU member state. We see huge press conferences when a company such as Siemens decides to open a factory here and government ministers are wheeled out to take the credit: it makes headline news. Yet such jobs have cost the taxpayer money in incentives to the companies to locate here: jobs may be lost in countries where they decide to close plant — as they may do here in future. There is no guarantee that the profits created by that particular group of workers will remain in the country, let alone the region: moreover local people may not buy the goods being produced, which are likely to be exported by road, raising congestion and pollution levels. If circumstances change — if say Turkey or Poland joins the EU, wage costs are found to be lower there and other circumstances are favourable — those companies could move to those countries.

The power of companies

I find it insane that any government is willing to hand so much economic power to companies, particularly when, like the Conservative government, it quite rightly recognises the issue of the Single European Currency as being a constitutional one, yet will let transnational companies have so much influence. Governments encourage the money markets too, without acknowledging the role they play in government. I also find it insane that the need to reduce unemployment and create jobs is so strong that we will pay millions of pounds to foreign companies to provide jobs here, rather than looking at more sustainable alternatives. It appears that the number of jobs is more important than whether they will last.

We are now seeing wage levels in Wales at the same level as in South Korea, although purchasing power is different. This shows how wage levels are driven down as a deliberate measure of competition policy, in the search for employers willing to come here.

We also see government being willing to hand the direct power to provide for a number of basic needs to private companies — water and energy supply being just two examples. Private sector funding is also sought in order to invest in public transport (or perhaps that should be — private transport for the public) and social housing, and this is not just Conservative Party policy. Why should the private sector be interested? Should profits be made from basic services at all and if they are to be made, where should that money go? We have seen enormous public anger at the pay levels of chairmen and directors of gas and water companies and the lack of power small shareholders have to change that. It has taken government action to deal with the issue of share options for such people, not shareholder power: such power exists more in the power to sell than in voting rights.

The question is a particularly vexed one in the UK, where the expected rate of return on such investments is much higher than, say, in Japan or Switzerland. More companies are floated on the UK stockmarket than on, for example, the German one. The money is also likely to be invested for a comparatively short period of time in order to give a rapid return. It is indicative of the short-term thinking which besets our politics and decision-making in general. Even church and pension funds are restricted in their possibilities to invest ethically and, in the case of pension funds, to invest in funds which accord with the interests of their own industrial background.

We are also seeing the government transferring the responsibility for provision for our old age to the private sector. This started when we were urged to take out private pension plans and opt out of the State Earnings Related Pension Scheme (SERPS): the pressure is on again to increase our insurance provision for our old age, so that we can afford the accommodation of our choice and our needs in later years. Why does the government have such faith in the private sector to deliver on this? Why does the government have so little faith in itself? What is to happen if the gambles of the private sector don't pay off? Does the private sector want that sort of responsibility? What is to happen if there are so many changes to the context within which our economy operates that money cannot be made in future as it is now? How can the private sector, with demands for a 20% annual return, be cheaper than the public sector with required rates of return of 5% – 8%?

I have yet to hear these questions asked in parliament. The answers would be fascinating. It really does seem that our politicians put all their faith in the power of money. They have totally lost sight of the fact that money is a symbol, not the reality.

I am reminded of a story I heard about the appalling inflation in Germany in the 1920s: someone put a large basket packed with the almost worthless currency down in the street for a moment in order to pick up something else. When they turned back, the money was still there, but the basket had gone: the basket had real value.

Power without safeguards

Money is now the largest traded commodity in the world. People become rich and poor through the speculations of the money markets. Black Wednesday and its effect on our economy was the result of financial speculation. That is real power and our politicians encourage it.

GATT has opened the doors for more of it. The Uruguay Round of GATT (which the British parliament agreed it was essential to bring to a rapid conclusion) stripped away many of the rights of governments across the world to protect their economies and their currencies from outside ownership and speculation. Governments have given up many of their powers in this area, but parliament still argues over the Single European Currency and ignores this wider erosion.

The economy has gone global, but the safeguards have not.

The only political movement in Western Europe to oppose the GATT Agreement was the Greens. In the US we saw the environmental and trade union movements working together to oppose the North American Free Trade Agreement and GATT. The reasons were the loss of governmental control over the economy, which many saw in terms of the loss of jobs in the old industrialised countries to the emerging economies with lower standards of worker protection and wages; the 'social dumping' of dirty jobs to poor countries and the lack of any form of environmental or social safeguards written in to the GATT.

Experience had already shown that protection of the environment runs a very poor second to free trade under the old GATT. Of the few cases taken to the GATT Tribunal for resolution, six concerned the environment. In all six cases, the protection rules were deemed to be a barrier to free trade and were overridden. So, just as Governments were signing up to protect the planet at Rio in 1992 at the Earth Summit, they were negotiating its more rapid use and destruction through GATT. While they were signing up to protect biodiversity, they were introducing intellectual property rights under GATT, which will allow companies to 'own' and patent bits of that biodiversity.

This intellectual mess is summed up in Article 2 of the Maastricht Treaty, which commits the Union to:

'... promoting sustainable and non-inflationary growth respecting the environment...'

We want to look after the environment, but carry on as we are. Greens say that cannot be done.

Our current economic practices are not delivering for the poorest and are creating increasing insecurity for many more. Yet, the UK is one of the richest countries in the world. If we cannot provide for our people *and* protect our environment, how rich do we have to be before we can do that? What are the chances for other, poorer countries in the current economic system?

To meet our basic needs, we must change our thinking about our future. We have to recognise the new challenges and find a way to meet them that secures our future — because it is not just our economy which is failing.

·4·
Failing on community

When people talk of how there used to be a 'real sense of community here', they will refer to being able to leave their door unlocked, to letting their kids play in the street, to having neighbours who would drop in or help you out — the sort of security indicators it is difficult to measure. It is the sort of life you will be told now only exists in 'Neighbours' or 'The Archers'. It is a community grouping that covers a range of ages and characters and, while there may be changes in membership and external factors (like employers), there is nevertheless a sense of continuity, together with a sense of shared values and shared space, which most people want to protect and where people cooperate with and support one another.

Now people feel that this sense of togetherness is disappearing and that communities are under threat. People no longer wish to be involved and prefer to remain anonymous. Some of the most shocking newspaper reports, I feel, are those where you can read of someone being attacked and in obvious distress and yet people walk past or turn away, not offering help or responding to pleas for it.

How can we maintain or develop any sense of community if we fear such basic involvement?

Fear of crime is certainly a huge, present-day problem. It is a general one which often bears no relationship to the risks we really face. Yes, there are places where it is dangerous to go out or even to stay in, but so many people feel that this applies to their street, their village, their town. Many women will refuse to go out after dark for fear that they will be attacked by a stranger, yet most crimes of rape or other physical violence are perpetrated by men known to their victim.

Our children are driven to school because we fear they will be abducted or assaulted — we teach them that cars mean safety — yet they are more likely to be injured or killed in a traffic accident or to suffer from asthma than to be abducted by a stranger.

We fortify our houses — bars, bolts, chains and alarms — and then leave them empty for most of the day as we move to other areas for our activities. If we don't know our neighbours and the rhythm of our neighbourhood, we do not recognise the abnormal and so it becomes more difficult to take care of each other.

The fear of what might happen reduces our lives and reduces many of our social contacts. We stay in, and the streets become less safe because there are fewer people on them. We get in our cars to feel safer and then begin to worry we might become victims of others' road-rage or we may be hi-jacked or have our car stolen or damaged. We are now beginning to see stories that women are afraid to drive at night, where once we saw the car as offering us protection and freedom from the fear of poorly-lit stations and threatening men on public transport. Those of us who walk, worry about the car that slows down beside us.

So our politicians offer us the promise that they will be 'tough on crime' and threaten tougher jail sentences, whatever the evidence about how effective they are. We have not yet reached the Californian situation where more is spent on the prison service than on education, but we are working on it: we already have a higher percentage of our population in prison than any other EU state. The fact that many of them are young, male and from working class backgrounds and that a disproportionate number of them are black (the groups with higher unemployment and poorer educational achievement) does not seem to tell the government anything about possible links and therefore possible remedies. It does tell us, however, a lot about what the establishment sees as 'real' crime: only a small minority are in prison for 'white-collar' crime, or crimes of polluting their workforce or the public — which can have appalling consequences for the victims, but maybe don't make such dramatic pictures for the news.

However, much though we want to see criminals caught (and many are not), we would rather not see the crimes committed in the first place and that means tackling causes in general — in so far as we understand them — and tackling criminals as individuals.

So, we stay in and make our homes mini-entertainment centres — one per occupant if we can afford it, so we do not have to watch the same television programmes, see the same video, listen to the same music. We will, however, worry about what children see on television. We no longer have to talk to friends and relatives directly by phone — we can contact them on the Internet. If we so choose, and have the money, we can reduce face-to-face contact and shared experiences at home, and elsewhere, to an absolute minimum. Some would say this can also be done by becoming old. Many households no longer share a meal together and have substituted no other means of being together as a unit.

We risk making almost every aspect of our lives 'private'. I believe this has serious implications for how we are governed and the sorts of decisions we make about our future. One of the few compulsory areas of 'antidote' to this is education — but even there, problems are arising.

A strong development in education over the last thirty years has been in the area of group and team work — off the sports field! Children have been encouraged to work together at times on a particular task or common project. Drama was a developing, and usually popular, part of the curriculum. A number of schools developed school councils, where certain issues could be raised and solutions sought. Many secondary schools had a pupil representative on the Board of Governors. There has also been a strong move to integrate more pupils with physical impairments into mainstream schools. We have also seen the development of strong policies in many schools to deal with and work to eradicate racism, sexism and bullying — shifting power relationships.

In these ways, pupils have been encouraged to develop a sense of awareness of the needs of others and to acquire the skills of co-operation and participation, necessary for effective problem solving. These methods have also been used to develop a more exploratory form of education — with an emphasis on forming skills rather than receiving information. Discussion, verbal report-backs, role play — these verbal skills were gaining in importance. Industry was telling schools that it needed people who could work independently and as part of a team; who could find information or ask for it as appropriate; who were confident verbally as well as in writing. The Technical and Vocational Education Initiative (TVEI) came into being, to encourage pupils of all ability ranges to develop in more ways than the purely academic. The new General Certificate of Secondary Education (GCSE) came into being — designed to widen the range of pupils sitting the exam — which, in a number of subjects, allowed for all or most marks to be granted for coursework, recognising that many people perform better if given time to think and consider their work and to reward those who work consistently.

The narrowing of education

Now we are seeing a gradual return to a narrower form of education and to an emphasis on personal and measurable achievement. Team spirit is to be developed in competitive sports on the playing field (if the school has not sold it to pay for repairs, teachers and so on...), whereas the growing emphasis has hitherto been on developing fitness through a wide range of sporting activities from the individual (such as weights, aerobics, swimming) to the team. How ironic that a government which, in theory, champions individual choice, has sought such uniformity for our children.

Classroom team methods are increasingly attacked as 'cheating our children', as if individual, 'private' work is all that counts in our lives.

The curriculum, too, is becoming more and more prescriptive in the search for the magic, uniform formula which will 'raise standards', without really addressing the question, 'Whose standards?' Secondary schools may now devote one day in five to the arts, economics, child development, home economics, non-European languages, philosophy, business studies, motor-mechanics, sociology, sport and other assorted and interesting areas which many of us believe help make us fully rounded people.

While there are some arguments for a basic curriculum which equips people with the tools for a lifetime of learning, the arguments for regular testing, which tells you what you know, rather than assessment which tells you how to improve, are much less strong. Fine, I suppose, if you want a nice set of simple numbers to tell you how you are doing — a sort of GEP (Gross Educational Product) which is limited in scope and unable to give any real qualitative information.

We are already seeing children 'coached for the tests' and teachers anxiously trying to deliver the curriculum regardless of local circumstances, such as classes of over thirty or children coming in at reception level from appallingly deprived backgrounds who have no real concept of what a pencil is for. Regardless too of the time requirements for pupils to really understand and explore topics and concepts. What is suffering tends to be context and cross-curricular links and the subject areas which are supposed to appear across many topics (such the environment or citizenship) which, because they belong everywhere may belong nowhere or, just as bad, become repetitive and dull.

Why do our politicians want to prescribe and measure? Many genuinely want all children to be literate and numerate, although their commitment to providing resources and effective training may be open to doubt. Where I really part company from them is on *why* they want our children to acquire these skills.

The world is full of beautiful and wonderful things — such as nature, art, literature, people — and faces many terrible problems such as environmental destruction, insecure employment, and social and personal deprivation. We need the personal resources to deal with so many things in our lives and on so many levels, from the deeply personal to the physical and the social. Our time in formal education should help us to feel more confident about dealing with a changing world.

Increasingly, education is really about equipping us to deal only with the paid-work dimension of our lives. Education is about training, we are told, in order to make our economy strong and to become internationally competitive. So we are told to develop our skills so that we can earn more for our country. All the three big parties tell us this and I find it soul-destroying, Our lives are so much more than work: indeed, many of us will probably spend considerable amounts of time out of paid work. What personal resources are we being encouraged to develop to deal with that?

Schools do, however, maintain something of their role as giving a geographical location to a community. They are still, particularly at primary level, a place where people from the locality can meet and be part of a local network. Even though parents delivering and collecting by car will often sit in their vehicles (engine churning out pollutants, making the roads congested and dangerous for the small percentage of children who walk to school) and not set foot in the playground.

In many other respects, our lives have tended to become increasingly zoned. This will vary from place to place, but in many respects we have seen the concept of 'community' designed out of our lives as different aspects have become more specialised. We can see this in the development of many large housing estates, whether publicly or privately conceived. They have often been built at the edges of towns or cities on green-field sites or derelict land. These were areas without a public transport infrastructure and, while roads may have been designed in, other facilities frequently were not. So if you had no private transport, it was likely you would find it difficult to get off the estate.

On many estates, no provision was made for meeting places. At best, there would be a small, single parade of shops, maybe a small community centre and perhaps a playground for young children — which would become unusable as older children gathered there and intimidated youngsters.

Even the better designed estates might add only a health-centre, a junior school and a pub.

If you had work, you had to travel to it — which frequently meant buying a car, which was then used by only one member of the household. To shop required travelling off the estate as did most social activities. The local shops were often expensive and/or did insufficient trade to stay open and were boarded up — adding to the sense of community inadequacy. If you had no work, you could hang about at home or outside in the little public space — just as teenagers in villages hang out at the bus shelter. Seeing people loitering in public then becomes intimidating to others in the area, so they do not go out. Young people, without their own transport or a decent, cheap local transport system, become dependent on their parents for transport or, if their parents don't run a car, become stranded on the estates.

Moving in the wrong direction

This is not just the picture in high-rise estates, there have been many low-rise, house and garden developments with the same faults, but at least there are the gardens to sit in. And still estates are being designed like this, on the sites of past NHS hospitals, on reclaimed land-fill sites, on unwanted farmland in villages. Places with no communal spaces, no playgrounds, community centres, shops, schools, pubs, libraries, places of worship, workshops... We still tend to build as if houses alone can create a community and omit, or find we cannot afford, the amenities which help to bind people together in a common sense of purpose and which can help to structure a vibrant community.

We now have purpose-built, distant shopping zones where we go for the primary purpose of consuming, unlike the High Street, which used to be a short walk or bus ride away for most of us, and which had (or has — if you are lucky) a multi-purpose nature.

People's living patterns have changed. Many of us no longer have time to shop daily for fresh produce and refrigeration has meant we no longer need to, so the trend is to shop on a weekly basis and take a car to carry it all home. The new shopping developments have tended to be on out-of-town sites, where parking is easier: these are often off major roads and require new slip roads.

They are single-function developments and generally privately owned. They have security guards and not the civil police force. They do not like people to loiter in their 'streets', talking, drinking, busking. They offer commercial, not social or democratic, space and will often refuse to allow groups to set up stalls to campaign on particular issues or people to collect for charities. While the video cameras and security guards can offer a feeling of security, these centres also offer us a restricted and sanitised experience.

Our lives have become compartmentalised so we come to do different things in different places. To do this, we rely heavily on more transport, whether public or private, and the busier and more congested roads divide us further and our local facilities wither. Because so many people commute to work, shops near our homes have less daytime custom. Because we are willing to travel to out-of-town shopping centres, local shops find they are unable to compete in terms of purchasing power and so tend to stock more expensive brand-name products and a more limited range.

Those who have no access to private transport, such as the poor and the elderly, end up paying more for their shopping or facing longer and more difficult journeys as their local facilities close. The 'community' divides still further. The problems are particularly acute in many country areas.

The trend toward specialisation and concentration is also to be found in the Health Service. Over the last few years we have seen the closure, or change of use, of what were local community hospitals, and the development of the regional super-hospital, concentrating many services on one site.

There are a number of reasons for this — not just government dogma. The changes have generally taken place in the name of efficiency and patient care: but 'care' in its medical aspect only, not care of the person as a whole. To give people a longer and often more difficult and expensive journey while they may be feeling unwell, and to overlook the difficulties this may present for those with dependents for whom they will have to make special arrangements, is not 'person-care', but it does help to externalise costs. The patients and their visitors bear those costs; the NHS no longer has to pay travel costs for consultant staff and keep a geographical spread of centres going.

The technological development of medicine has also had a centralising thrust. As equipment becomes more expensive, it is more costly to equip a number of centres: to get maximum value for considerable expenditure you want the maximum use for such machinery and it is logical to bring patients to one centre. The expertise then concentrates as well. The same argument applies to more complex surgical procedures and other processes: it is logical to have the expert team in one place and to bring patients to it. Improved understanding of medical care in, for example, road accidents and other trauma cases is leading to larger trauma centres, often resulting in the closure of more local Accident and Emergency Departments.

The former community hospital either closes — and is sold off as a private nursing home or to be redeveloped to provide revenue for the Trust — or becomes a specialist centre for the left-over, non-intensive services such as geriatric care or minor day-surgery procedures. Care has been transferred to 'the community', which may no longer exist.

The concentration of power

The trend towards centralisation and specialisation, whatever the motivation, can be immensely damaging.

It removes the feeling of ownership from local people, who have no sense of control over, or responsibility for, the ways in which their lives are run and decided. We feel that things are being done *to* us, not *for* us and certainly not *with* us.

We can exert power only through complaints and cash — we have consumer rights, but not citizens' rights.

We are seeing the development of the 'compensation' economy, where if anything goes wrong someone must be at fault and we should be paid for their mistakes. In many cases this is an advance — people's lives can be devastated by the negligence of others, so it is only fair to receive a sum that can make life more bearable for such people and their relatives. But I think we are moving towards a situation where we risk reducing the *value* of everything to its monetary *cost*. There is an ambivalent attitude which says on the one hand that money will make everything all right (the grown-up version of being kissed better by mummy), so if we pay you enough, we don't have to rectify the problem or even admit liability: on the other hand, we seem to feel that if we threaten to sue and demand enough cash our problem will be taken seriously and there will be a response. All too often that is true. We frequently hear people say after court cases: 'It wasn't for the money. We wanted the truth and we want to stop this ever happening to anyone else.' So often, the authorities' fear of the truth, of the public having access to information, creates more problems than it solves.

We are being encouraged in this by the Citizens' Charter. This is fine, as far as it goes. I can see nothing wrong in people being told what standard of service they should be able to expect from a particular body and what to do if that service is not forthcoming. I do see problems in this if the staff providing the service are not involved in deciding what is feasible, and what are the appropriate resources to provide the service. The beneficiaries must also be consulted as to what they want.

But the Citizens' Charter should not be confused with any real advance in democracy. It is fundamentally a Consumers' Charter: it portrays us as receivers, not as participants. It was introduced to head off growing public concern about constitutional issues, such as state secrecy, poor government, and growing privatisation — a way of saying that government and public services are accountable, so no real reform is necessary.

The Citizens' Charter was billed as John Major's 'big idea' and then politicians wonder why the public is losing faith in them!

Government by quango

The worry about government accountability has grown alongside the development of the 'quango' state — the privatisation of government itself. Government, whether at local or national level, has always found it useful to pass the responsibility for administering certain services to an appointed body, which is then held responsible to a particular Minister or committee. It has been used as a way to reduce the workload of elected members and involve more people (the so-called 'great and the good') in some form of public service by using their skills or, some might say, rewarding them for their political support by a position of some authority.

Over the last few years, however, we have seen the transfer of powers from *elected* bodies to *appointed* bodies and a considerable increase in the numbers and powers of the latter. We have seen the fragmentation of power to bodies whose roles are often not fully known about or understood: many people feel they no longer know where to go if they want information or have a problem. There is often no requirement for public access to meetings, papers or decisions of the bodies. Many are responsible to a Secretary of State who, in theory, is then answerable to parliament for their actions: all too often it leads to difficult issues not being dealt with in public. We have only to look at the public concern surrounding Health Authorities to see this exemplified.

This system leads to increasing powers of patronage within the government and growing public distrust. While John Redwood was Secretary of State for Wales, for example, it was noted that the majority of the chairmen (sic) of the bodies appointed by the Secretary of State were Tory Party supporters. This was hardly likely to develop the trust of the Welsh public, less than three in ten of whom vote Tory.

It is also a way of centralising government while claiming to 'get government of the backs of the people' — to use the campaign slogan of Margaret Thatcher on 1979. What it does is to give power to the executive (the Cabinet), by allowing Ministers to rule through Regulations (which only have to be tabled in parliament but not voted on) and through residuary powers. We have seen this happen in education, when Gillian Shepherd (Secretary of State at the time) used her powers to over-rule Hackney Borough Council and remove the sitting governors at Hackney Downs School in order to send in her own appointed expert team to solve the school's problems: they decided to close it.

In 1994, the Democratic Audit estimated that quangos were responsible for spending some £46.6 billion, nearly one third of total public spending. However, the Audit's definition of quangos was very wide-ranging.

At the root of this form of government is a belief that people do not really mind how a service is delivered or decisions reached as long as it happens. This belief sees the public as consumers, wanting 'the goods' for the lowest price. This would be reasonable if we really did have a genuine choice, if public, private and voluntary sectors were on a par, the mythical 'level playing field' of politics. But this is not the case. Compulsory Competitive Tendering, for example, forces local councils to accept the lowest price for the job and to exclude any other considerations in their choice of contractor, such as giving work to locally-based firms, those with an ethical purchasing policy, or even those with a reputation for, and commitment to, quality.

Such a belief also fails to see that most people are both providers of services as well as consumers, and as providers will want things such as reasonable pay, conditions of service and the resources to do the job and respect for a job well done. We also pay for services and wish to have a say in what is being done with our money.

We may want a say in the ethos and principles of what is provided, which cannot be measured in cash alone. We are people who can think and make moral judgements — not just one-dimensional beings worried only about cheapness. Being able to exercise those judgements should be a democratic freedom.

Britain prides itself on being a democratic country: we like to feel we have the 'Mother of Parliaments' and that we have been instrumental in bringing other countries to an understanding of what democracy is about. Much of the triumphalism when the Communist regimes of Eastern Europe began to crumble was about the values of freedom — others could have what we enjoy.

And we do enjoy many freedoms. Anyone who has spoken with dissidents from East Europe and parts of Africa, Asia and South America soon realises that there is a qualitative difference in our lives in terms of political freedom. I feel it is important to recognise that. In general we have the opportunity to campaign openly for change without the prospect of 'disappearing' or being executed for dissent. Living in Britain in the 1990s is not like 'living in Ceaucescu's Romania' as I once heard a councillor claim — we don't live in a totalitarian dictatorship. To minimise the differences degrades and belittles the suffering many have faced in their quest for what we tend to take for granted.

However, that is not to say that everything here is perfect. It is not. Possibly the greatest threats here to democracy are complacency and inertia — a feeling that we have 'got it about right'. It is an arrogant attitude that refuses to admit we might have something to learn from other countries.

Britain is the only EU state which does not have a modern written constitution: indeed, it is one of only two modern democracies in the world not to have such a thing (Israel is the other), although all former British colonies do. While a written constitution is only a democratic safeguard if it is respected and implemented, it nevertheless provides a yardstick against which legislation can be measured. By making parliament our supreme decision making body and going for a system which allows each parliament to legislate as it likes, it is possible to erode freedom and democracy without effective challenge.

A whole layer of government was swept away from Londoners and other big-city dwellers when it was decided to abolish the GLC and metropolitan counties without any consultation at all. Local government has been restructured three times since 1971 because different parties (or the same one) have different ideas at different times. Yet there has been little open debate about the role and powers of the different layers of government and there is no constitutional framework in the which a debate involving the public can take place.

So the government decides on the amount of money local councils are given and how much more they can raise formally from local people, and it can take action against councils which do not toe the government line. Government decides on the powers and duties of local government and can remove and impose these as it chooses. Local councils are then blamed (and are often legally liable) for not delivering effective care in the community, street cleaning or housing. Is it any wonder that only one in three people bother to vote in local council elections, when they know that the real decision making power is elsewhere?

The Criminal Justice Act is another example of the erosion of democracy and freedoms and is justified, of course, as 'protecting society'. It has qualified the centuries-old right to silence, made civil offences such as squatting into criminal offences, and has further restricted the right to free assembly. It was a panic, catch-all reaction to stamp on behaviour which was 'different' and it gave the police powers that they did not want and expanded their role as agents of the state rather than servants of the public. Instead of looking for sensible solutions for travellers with inadequate sites and people with no homes, the government decided to create new groups of criminals and thus shifted the issues from the area of society and community to the area of the courts and prisons.

One of the measures of democracy is how well it can cope with safeguarding minorities: it is not just about imposing the will of the majority. That is why a Bill of Rights is usually part of any constitutional settlement.

We have a very unsophisticated way of even determining the will of the people. We have an electoral system which does not reflect people's voting intentions (1987 saw an overall government majority of one hundred seats on 42% of the vote): it effectively tells many of us not to bother voting. The government elected under this disproportional system then claims a mandate — that is, public support — for its entire manifesto, which the vast majority of voters have never read. There are also many politicians who will say that elections are not about electing a government but are simply a way of getting rid of one you don't like. The voting evidence says that elections do not really do either effectively. Confusing isn't it?

So we are controlled by a centralised, erratic political system which does not really represent society (only 11% of MPs are women, for example). Local government may be more representative in some respects (there are more councillors from ethnic minorities) but is largely controlled by national government. This system is separating the different aspects of our lives so that we, like the system itself, become more fragmented. It sees people primarily in terms of their economic worth and ambitions rather than seeing us in an all-round, holistic way. Is it any wonder that such a system cannot create and conserve meaningful communities?

· 5 ·
Global insecurity

In 1989, there was great euphoria at the destruction of the Berlin Wall. The feelings of joy and hope as that potent symbol of repression and division was breached were exhilarating. Happy news filled our television screens for a change.

For Greens and others in the peace and environment movements it was particularly poignant. Many of those involved in working for change had done so through environmental groups. The East, like the West, had tended to view the environment as apolitical and generally as an area of scientific research rather than of political change. Those of us privileged to meet with the Greens from East Europe during that time of change were impressed by the enormous courage they had shown and by their determination to look for political solutions. They wanted solutions which did not involve embracing Western capitalism and consumer values at the same time as developing democratic and open societies.

We were told of the appalling environmental damage in their countries: land heavily polluted by the military; the poorly-maintained chemical factories built over the aquifers supplying drinking water to thousands of homes; the creaking nuclear power stations; the toxic waste dumps filled with the poisonous by-products of industries from both East and West.

When Members of the European Parliament's Green Group visited the 'new' democracies, these were the sort of sites they were shown. The people they met were often those directly affected and who had come to the strong conclusion that there had to be a better way — one which did not destroy the environment and the people but which could meet people's needs. This was why they had been prepared to question the state and then stand for election as Greens when they could.

Within the peace movement at the time there was real hope that the insanities of the arms race could be halted — if not reversed. With the collapse of the old bloc system, the 'enemy' had gone, so the rationale for spending an obscene amount of money on ever-more sophisticated ways of destroying each other should also go, it was argued. Already, the Strategic Arms Limitation Talks (SALT) had appeared to reduce the nuclear threat to our survival. At last, the 'Peace Dividend', the money released by reducing arms spending, could be spent on providing clean water, cash for health care and education, and a clean environment. That was the hope.

President Bush spoke of a 'New World Order', implying a bright, dynamic future as the old Soviet Bloc converted to democratic, capitalist economies — like the United States! Commentators spoke glibly of 'the end of history' as if we were all now on the same side, moving towards a Brave New World. Western, industrialised countries promised help and promptly rushed in their businessmen, management consultants and, in some cases, environmental and constitutional experts. There was a rush to cut deals over access to resources such as natural gas, oil and diamonds. Nuclear companies, losing their markets elsewhere, sought deals to assist the ramshackle nuclear industries of the emerging democracies, particularly in the matter of dealing with nuclear waste. 'They are looking for new dumping sites,' said the cynics.

Not surprisingly, the pace of change soon slowed. Changing a political culture requires not only change at the top but new ways of working and new goals throughout the administrative infrastructure in its widest sense. It became apparent that to restructure whole economies and support the people was an extremely expensive and difficult task — especially for 'free-market' Governments, committed to the minimising or removal of state support. How were they to justify supporting the transitional economies in Eastern Europe, when they were refusing to support domestic industries?

So, although we had seen the US government working for the overthrow of communism, stoking its own military force, weapons development and space programme to force the Soviet Union to distort its economy in order to maintain its superpower status, when the cracks finally appeared in the edifice, we saw the US and other Western Governments with no effective strategy to cope with the changes. It appeared as if the effort had gone into winning the Cold War, not into creating the peace.

It may be that events moved so rapidly that 'management' was not really possible. But it seems to me that the West has been fundamentally dishonest in selling a fairy-tale. The pain and pace of readjustment was undersold and support in assisting the democratic changes has been piecemeal. A number of problems which have surfaced had not been seriously considered by outside governments: property rights, for example. What happens to those farming state land, if that land is restored to the families of its former owners or sold off to private buyers? What happens to people's housing prospects when the prices of city dwellings escalates as foreign speculators move in? What are the implications for the development of a stable, democratic society if the state cannot afford to pay its police, teachers, administrators, military and scientific researchers? How can a currency maintain its domestic value when it is exposed to the international economy?

The instability and the vacuum in infrastructure, together with the resulting lack of hope, can open the way to other forms of 'order' — organised crime and authoritarian government. People want security and dignity and will often respect those who provide them. Instability is a fertile breeding ground for national fundamentalists, who are not interested in universal human rights but only the rights bestowed by blood, birth and language and thus treat others with contempt and violence. Where instability and hopelessness abound, crime often seems to be the only way to make money, both as the means to live and to have status. The power of money is a Western value soon understood and acquired.

Prostitution and drug-trafficking are two of the fastest growing parts of the economy, not only in Eastern Europe. Very few pay taxes on this income and thus society suffers without gaining very much. A Russian teacher or doctor can earn more as a stripper, a drug-pusher more than a police officer or water engineer.

Part of the rationale for the EU in setting up the European Bank for Reconstruction and Development was to help stabilise the economies of Eastern Europe, so that the EU would not be faced with thousands of economic migrants looking for work. It is apparently a good thing to be such a migrant provided you 'get on your bike' and move within your country, but not if you cross borders and you are poor. That fear of large-scale migration has been a crucial part of the thinking leading to the introduction of severe immigration and asylum rules at the EU level. The 'harmonisation' of policy is leading to a levelling down to the harshest standards and to the creation of what has become known as 'Fortress Europe' — difficult to enter, particularly if you are poor or non-white.

Maintaining the nuclear edifice

One of the other expanding aspects of East Europe's 'informal' economy has been the sale of weapons and expertise in their design and manufacture. The ex-Soviet Union had a massive nuclear and non-nuclear arsenal and had many scientists deployed in research and development as well as in the civil nuclear energy programme. Other governments simply failed to face up to the likely consequences of inaction. Western governments were urged to buy up the weapons and decommission them and employ or retrain the scientists and close the nuclear plants. If not, the argument went, these weapons would be sold off indiscriminately; the scientists would be 'bought up' by governments wanting to develop nuclear weaponry and plutonium would be sold to whoever could afford it.

Basically, Western governments chose instead to regard the weapons and the development programmes as an internal issue for the ex-Soviet Union. The nuclear power plants still operate and plutonium has been found in transit in Germany; the scientists are dispersing — many of them unpaid by the their former employers.

What was behind this lack of decisive action?

Many governments were trapped by their own political positions. It is perfectly acceptable, they argue, for them to have nuclear weapons: these are a legitimate deterrent for responsible administrations, and a matter of national sovereignty. It is also a measure of international status — the United Nations Security Council's permanent membership comprises the nuclear powers and it was felt essential to keep Russia as a member of that Council in order to help maintain stability and a Western-friendly government in that part of the world.

It was also felt that to buy the weapons would be too expensive in conventional economic terms and too complicated. It would also have enabled the peace movement to say that if the ex-Soviet Union no longer needed its nuclear weaponry, then the West could also disarm in this way. Allowing some of the new independent states to keep a nuclear capability has thus helped to retain the nuclear status quo — which represents an enormous investment by governments financially and in human terms.

In 1991, however, we saw the outbreak of the Gulf War. This provided another focus for the creation of an 'enemy', and meant that the retention of a huge military capability could be justified. It also provided a useful test-site under battle conditions for new weapons.

Many were stunned by the rapid about-face of many of the governments involved. These were administrations which had supported Saddam Hussein and Iraq in the long war with Iran and had supplied weapons to Iraq (some had also supplied weapons to Iran — presumably to maintain a military 'level playing-field'). This support had been maintained despite the public knowledge that Hussein's regime had gassed many of its Kurdish people, persecuted the Marsh Arab population in the south of the country and generally ran a brutal and repressive government, with Hussein himself aspiring to leadership of the Arab peoples. His government had been supported with a view to maintaining Western influence in the area, partly through habit and its position in relation to the Eastern Bloc and more particularly because of the oil reserves.

As the background to the war became clearer, it appeared to have resulted from a failure in international relations. Mixed messages had been conveyed to the Iraqi government: whether or not this was intentional is a matter of debate. It was a war, however, with huge human and ecological costs.

Many would say that the tragic situation in ex-Yugoslavia has also been created through neglect and mis-reading of the situation. Some governments supported the creation of separate nations without demanding a recognition of fundamental human rights. This creation was justified partly to permit the intervention of the UN, which can only intervene at the request of a nation state and not at the request of people within a state. The area was also used as a testing ground for EU agreement on foreign policy issues (introduced under Maastricht), before the logistics of that agreement had really been thought through and while the relationship to the UN and NATO was unclear. The US had also chosen to disengage itself from involvement in Europe, but had been unable to avoid directing from the sidelines. There has been a lack of clarity over the level and aims of involvement in this conflict. As usual, developing peace is proving a difficult and expensive task.

Much of this is due to the fact that more attention has been given to avoiding another world war, while maintaining an East-West balance, than in trying to prevent wars and create peace. There have been numerous, smaller-scale, protracted wars since 1945 in which the opposing sides have often been supported and encouraged by the two political blocs. An enormous amount of money has been made through these wars, and spheres of influence, and access to resources, have been developed as a result of them. The majority of the dead have been civilians.

Arms have been sold to horrifically repressive regimes because that suited political beliefs and economic policies. The nature of many of the products sold has demonstrated that many of them are for internal use and not external protection. That has not worried governments, as we have seen in the UK in the Pergau Dam and Iraqi arms sales scandals.

Our government excels at sending mixed messages to countries we wish to influence. Those messages tend to heighten tension, not just in Eastern Europe but globally.

The creation of poverty

Those tensions are thrown into even sharper relief as we come to understand that we are interdependent on this planet and that the national interest cannot be as narrowly defined as in the past. The United Nations Conference in 1992 on Environment and Development (the 'Rio Summit' as it is often referred to) made the connections between us absolutely clear. The commemorations of the fiftieth anniversary of the ending of World War 2 and the dropping of the atomic bombs, together with the French government's resumption of nuclear testing has reminded us, if we needed it, of our capacity to wreak destruction on a global scale. We know we are One World, but we are not clear how to deal with that knowledge.

We tell the world that there is only one way to develop and that is *our* way. Through the International Monetary Fund — where representatives from five of the world's industrialised countries (the UK is one of them) have more than 40% of the voting rights as countries vote according to the proportion of the money they put in — we push countries wishing to obtain money to enforce a disciplined spending regime at home. This means they have to cut public spending and expand their export trade in order to shift the ratio of income to borrowing. Cutting public spending means they have less money for health programmes, education, development of clean and reliable water supplies and so on. Yet resolutions passed at UN conferences and UN programmes are often aimed at countries doing just the things they are required to cut. There is a lot of research which demonstrates that the quality of life for all people in those countries is dramatically improved by the provision of such services.

To gain foreign currency to repay loans and provide higher personal income to some (the theory being that this will then permeate the economy and eventually benefit everyone through the purchasing of goods and services) exports are encouraged. Given the small manufacturing base of these countries, the exports tend to be natural resources or cash crops (crops raised for sale, such as coffee or peanuts). Thus we have seen acres of forest land laid bare in Brazil, Indonesia and other countries in order to sell the hardwoods not only for building and furniture but also for chipboard and cardboard. However, little investment was made in looking at the longer-term environmental and social effects of this destruction.

The habitats of many plant and animal species have disappeared, been severely depleted or are under threat. Water tables have been affected and soil stability has been eroded. For example, the catastrophic floods in Bangladesh over recent years are believed to have been made worse because deforestation in the Himalayas has meant water is no longer retained in that area but is free to cascade downhill unchecked. The change in water tables means local agricultural production is affected, fuel-wood and building materials are no longer available, so the local population either moves further into the former forest land or moves out altogether, putting greater pressure on urban centres or other farming areas.

The need to produce crops for export has also resulted in land being purchased or forcibly taken from the local population, who are again pushed out onto more marginal land or leave for the cities.

However, as more of such crops come onto the market, the prices tend to fall, so income falls also — then more has to be produced simply to make up the shortfall in income and the downward price spiral begins. It is a dispiriting, no-win situation. Is it any wonder that exploitative, soul-destroying alternatives arise?

In some countries we see the expansion of the production of cannabis and opium poppies in order to supply the ever-expanding markets overseas. This can provide enormous sources of foreign currency, but not usually to the government through formal channels. If anything, it often provides a rival power base and politicians and governments are bought and intimidated by crime money, with the West (usually the US) spending large sums in an unproductive effort to stem the tide. I find myself wondering if those who set the rules for the economic system that has encased the world's developing countries and forced them to find markets for whatever products they can, can ever see themselves as responsible, in part, for the growth of the drugs industry? And if so, would they change those rules and values?

Would they look to export substitution programmes, or would that be contrary to free trade?

We are seeing a growing repugnance at the use of child labour in many of the world's poorest countries. In Pakistan, for example, some of those who are fighting against the use of children in the carpet making industry have been taken into custody and face charges of sedition, which carry the death penalty. Such is the importance of keeping labour costs down and foreign exports up, that to campaign against the use of young children working long hours in poor conditions is seen as acting against the interests of the national economy and thus the state itself. We have long known of the appalling child sex industry in Thailand, but we also know that it brings an enormous amount of foreign currency into the country and that many people there claim that their families are only kept alive because selling children gives them an income.

We know that many people leave their homes and work in another country in order to have money to send home to keep their families and maybe provide their children with an education. We know that some of these women working abroad live in virtual slavery and are barbarously treated — we have seen such cases in the UK. We know that foreign workers are often poorly paid and lack many basic civic rights (such as the Turkish 'guestworkers' in Germany or the Vietnamese 'giftworkers' in communist Poland). We also know that in times of trouble they may be expelled and forfeit all their savings and possessions, as happened to many Pakistani, Bangladeshi and Palestinian workers in the Gulf during the war there. The foreign currency that such workers send home often makes up a substantial proportion of their native country's income.

Many are rightly appalled that people can only make a living by leaving their family and home country — this is not the 'free movement of people' that is part of the European vision. This is forced migration.

But it can be difficult for poor countries to develop their own domestic industries. The 'added-value' to raw materials that comes through manufacture or processing is often not carried out in the country of the product's origin or, if it is, the plant may be owned by a foreign company so the profits are exported as well as the goods. Why?

Sometimes, the originating country lacks capital to invest in production but it can also be due to the barriers which often face poorer countries wishing to export to already industrialised countries. These have increased substantially in recent years: the Overseas Development Council estimates that such barriers have cost developing countries between $10 and $15 billion a year in textile and clothing exports alone. The EU, for example, operates a range of non-tariff and tariff barriers, which rise according to the degree of 'finish' a product has: it is therefore cheaper to import raw materials and process them within the EU boundaries and keep the jobs there. That is one of the reasons that Japanese car manufacturers have been so keen to set up factories in the EU — to have European production and thus beat tariff barriers. Countries such as Sudan, Somalia and Bangladesh unfortunately do not have the financial muscle to do the same.

In these circumstances, it becomes impossible for the poorest countries to get off the bottom rung. They have done what the IMF and many financial advisers and foreign governments have urged them to do — cut public spending and increase foreign revenue raising activities, including the export of precious natural resources. The outcome is increased poverty, social disruption and environmental damage.

Given that many countries still carry a heavy debt burden to banks and other financial institutions — anxious to lend during their boom years of expanding oil revenue — it is no wonder that they are in poor financial shape. Simply servicing the debt (an estimated $1.5 trillion in 1991, including the countries of Eastern Europe) costs billion of pounds each year. It is estimated that in 1992 for every $1 received in aid, $2.5 was paid in simply servicing debt. In 1991, a staggering 87% of Uganda's export revenue was used for that purpose; the figure for Brazil was 45%. While various schemes have been introduced to reduce the debt burden of the poorest countries, the damage has been such that there has been little real difference. Many nations of the world have seen their economy distorted, with all the knock-on effects, because they were urged to borrow over their heads.

These countries have been pushed towards the global free market economy and many have signed up to GATT, thus opening their financial and other markets to the industries of the world. But they are not equipped to play the same game: less training, investment and practice means this is not an equal match. They are not in control of their destiny.

Population and over-consumption

And yet there are still those who believe the problems of the world's poor to be self-inflicted. 'They have too many children,' we are told. Yes, it is true that the population of many poor countries is growing rapidly and more young people may lead to more growth, but the issue is not simply one of numbers. In places where infant mortality rates are high, it is not surprising that women give birth to many children, simply so that some may survive. In cultures where the state does not provide for the old, children become important as a means of support. Lack of access to effective birth control or cultural resistance to it, may also mean women have more children than they want. However, people's ability to bring up healthy, educated children and provide them with a decent standard of living is not always within their control. The displacement of people from their land whether for example to grow cash crops on it, or to construct a dam, is not under their control. If their soil is eroded, because they are being forced to grow unsuitable crops; if their water supply dries up because industry needs more water; or their land is polluted by the industrial activities of an oil company — that is not the choice of the people. They have become victims.

So the population question is not merely one of numbers but of resources. The industrialised countries, liking to feel they have no problem because they can feed their children, play their part. We feed our children because we are dependent on other countries for much of our food and economic income. Our governments devise the rules of the IMF, the World Trade Organisation and the European Union. The governments we elect help to keep many of the world's people poor. The miracle is that so many children survive, when they have so little.

There is a growing problem in matching resources to people and securing a decent standard of living for all, not just some. What that standard of living should — or can — be is problematic too. But it is a consumption issue as much as a population one: the two are inseparable.

The world's poor are also described as 'backward' or 'primitive', meaning that they have not followed the path of industrial development, as if this is the only route worth taking. Such a view also implies that a culture based on the written word is superior to one based on an oral tradition, which is not the case. Both can function effectively, though differently. Much of the communication in the industrialised world is a result of the fact that we are a dispersed society, a consequence of our agricultural and industrial history. Our society is complex and heavily dependent on the written word: illiteracy often results in exclusion from paid work and in social disadvantage. As a result of changed trading patterns and the development of the global economy, literacy is becoming an increasingly important factor in self-determination. To oppose the building of a dam on your land, or the activities of a mining or logging company, requires literacy, if only as a means to know your enemy. There is also a considerable body of evidence that when women are literate they are more likely to have the number of children they want. Literacy assists independence.

We need to question the motives of those who wish to cut spending on education, health care or birth control programmes. Are they really interested in democracy, people understanding and having power in the decision-making process — or do they prefer people to be too bound up in the struggle for survival to have time, energy and the means to be involved?

However, we are now beginning to see a significant change in the global balance of power. Successful economies, in current thinking, are those which have highly skilled workforces: human intelligence is a valuable commodity and adaptability is essential. The new economic threat to Western hegemony is found in the so-called 'tiger economies' of the Pacific Rim — virtually by-passed by the Industrial Revolution.

But a deeper threat is posed by the decay and destruction of our global ecology. We have only recently begun to realise the intrinsic value of vast forests to our weather patterns and the gene pool. Whereas the world's industrialised countries have, in the past, wanted the natural resources of South America, Asia and Africa as cheap imports, they are now looking to some of these countries as economically threatening on our own terms, having followed our own economic ideas and the development path we took ourselves. We are now slowly realising that we might have sown the seeds of our own destruction.

· 6 ·

Ecological insecurity

Our political system is failing the environment because it doesn't take it seriously. Despite all the campaigns, all the research and reports and all the good intentions in so many speeches, our environment is not seen as an inclusive part of the dominant political philosophy. Part of the reason, of course, is that so much of the work to do with the environment in terms of landscape, farming and how we look after what Brundtland described as 'the global commons' has been, in Britain, the preserve of pressure groups and not the preserve of political parties. Parties have relied for their policy development, and for being made aware of whether something was a growing political issue or not, on outside activity rather than their own internal research. British politicians obsessively see the economy, and maybe defence, as the be-all and end-all of their job. These are the grand issues of politics, traditionally the preserves of men, and the environment has long been viewed as something which is left for people who are not formal politicians. The Department of the Environment has historically had more to do with local government and housing than with many environmental issues.

We have a very long history in this country of pressure groups being concerned with issues which are now seen as part of Green politics. We have the RSPCA for over 100 years acting in pursuit of animal welfare and increasingly animal rights: we have the National Trust which has looked at how we preserve landscape and countryside for ourselves and our general use. Politicians have been content to see the land as a source of power and wealth; in this view land use becomes a means of increasing economic well-being. This has a paternalistic side to it — a notion that you need great landowners with a sense of vision to look after the countryside.

Increasingly, too, we have become displaced from the source of what sustains us and have been moved into the cities, a much more constructed and planned environment where our natural resources have been used to house, heat and transport us. Yet we are actually removed from the production of those resources.

The vast majority of decision makers come from backgrounds which have nothing to do with supplying our basic needs. If you look at the employment background of our MPs you will find that they come from the law, from academia, and from certain branches of business: the number of farmers is actually very few, a considerable change from the past. Few come from the industrial shopfloor or from mining, fishing, or energy production. So those making the decisions are those who live in an artificially constructed environment, as do most of their constituents. As a result we see the countryside not as a place which supplies food and energy but as a visual experience which we visit: we don't see it as a working environment on which we all depend.

Our system, like so many others, is also focused on the idea of defending the national interest, both in a geographical and an economic sense: politicians must put *the nation* first. I once heard a Conservative MP define the key tasks of her Party as securing 'a sound currency and defence of the realm.'

So what we have in British politics at the moment is a growing sense that the environment is in some way important, that the public considers it to be important, but yet the tools and the history which politicians have at their disposal do not allow them to recognise the need for action on environmental issues and implement the necessary measures with ease.

There are tensions growing within the large parties in the UK about where the environment in its widest sense fits with their politics. In 1992, the UK government attended the Rio Summit and it signed up to the agreements on such issues as biodiversity and global warming. It has put money into research, it is recognised as having done something positive to promote Local Agenda 21, and it has published an environmental stock-taking report which although criticised, was produced — which is more than can be said for a number of governments. John Major did not go to Rio with his arm twisted behind his back, as did President Bush, and we didn't hear our Prime Minister making the sort of crass statements which we heard from Bush about not allowing protection for the environment to get in the way of American jobs...

Bush's view exemplifies the current political perspective: jobs and the national economy are important, the health of the planet which supports us is not. It's the old: 'Trees don't vote, people do.' As if people are not part of nature.

Will matters get worse?

The environment — air, water, earth, natural resources, flora and fauna — is seen as a given: the ultimate in free consumer goods. Because it is viewed as infinite, something which has always been there and therefore will always be there in future, we do not have to take care of it.

For many, this is a view bolstered by their particular religious faith —

God, or whatever name you use, gave humans the earth and everything upon it for their use and has promised not to destroy creation. So, as we have divine insurance, everything will actually be all right — at least for those believing in the right things, who will presumably be safe from nuclear fallout or depletion of the ozone layer. Fortunately, from a Green perspective, not all religions or believers think that way: there are those who see our world as something given in trust to us, for which we have a duty of care.

Others see the planet as a vast, self-regulating system, with a capacity to heal and maintain itself. Whether humans will survive, or for how long, as part of the cure is not clear.

Yet others see scientific invention through human creativity as the key to any problems we might have: scientists will find a way to counteract global warming, deal with nuclear waste, save the ozone layer and produce a pollution free, fully recyclable private car before things get really bad! Meanwhile the British government still puts more research and development money into military expenditure than anything else; the emphasis on research in our higher education is to make it relevant to the market and the West has come up with no constructive, overall plan to employ or retrain East European scientists, many of whom were involved in weapons development. This is not a sensible use of resources, nor grounds for scientific optimism.

In the meantime, we are seeing our natural systems become increasingly overloaded, whether through pollution, exhaustion or direct destruction. Many systems are close to collapse or changed beyond easy remedy. A growing number of people believe we are treating our common support systems of air and water in a non-sustainable way. Recent opinion polls show a majority of people believing that the future will be worse than the present.

What is the evidence for that point of view?

Two important areas where we can see a systems overload are in the erosion of the ozone layer and the evidence of global warming.

The ozone layer is a protective stratospheric layer between the sun and the earth. It filters the sun's ultraviolet radiation and thus softens its effect. Losses from the ozone layer were first noticed over the Antarctic in 1975, and have continued to grow since then over other parts of the world too.

The losses are primarily due to CFCs (chlorofluorocarbons) — gases which, due to their very nature, reach the stratosphere unchanged. There they react with the ultraviolet radiation and remove the protective barrier. CFCs have been used in refrigeration and air conditioning coolants, in aerosols as a propellant, in the manufacture of some forms of packaging, and are even a by-product of space travel. The problem arises when the gas escapes.

Ozone depletion matters because without that protection many changes occur. In the Southern hemisphere in particular, there has already been a dramatic increase in the number of skin cancers and eye problems — in other animals as well as humans. In the UK, skin cancer is now the second most common kind of cancer. There is likely to be an effect on the food chain, as increased ultraviolet radiation affects the growth of phytoplankton — the basis of the marine food chain. Many current crop plants react badly to an increase in such radiation, with the possibility of reduced food production.

While drastic international action taken through the Montreal Protocol to phase out the production and use of CFCs and other ozone-depleting chemicals has achieved considerable success (dropping some 60% from their peak in 1988), the effect is not easily remedied. The effect of those gases already released has not yet totally worked through; nor are all such gases out of use. Commentators believe the additional effect of growing refrigeration use in the developing countries will be a considerable problem.

We may yet have to pay the full price for overloading the system, expecting it to deal with whatever we choose to throw at it.

The same is undoubtedly true of climate change or global warming — the so-called 'greenhouse effect'. By increasing the amount of certain gases emitted into our atmosphere we have increased its temperature, so more heat is now held in the atmosphere than even a century ago. This has upset the balance, with potentially enormous and disastrous effects. There are still some who dispute that the greenhouse effect exists or that it is anything more than a periodic shift in the earth's temperature — it's happened before, the sceptics tell us, and we're still here.

What the sceptics don't take into account is that changes in the earth's temperature in the past have meant famine and water shortages in some places and the extinction of various species. That happened when the world was a less populated place in human terms, when trade and national economies were not inter-linked as now and when humanity was not aiding its own destruction as enthusiastically as it is at present.

We cannot 'wait and see'

A policy of 'wait and see' is really not a viable option. Our food crops are acclimatised to certain growing conditions — a change in growing temperature can mean lower crop yields. A shift in the world's main growing areas may not yield sufficient fertile land to grow enough crops. On current predictions, we are not likely to see a smooth transition to different conditions, but are likely to suffer unstable and extreme weather conditions — again a threat to crop production, let alone any associated damage from storms, hurricane or drought. With a rise in temperature comes a change in the distribution of insects and associated diseases, probably hitting vulnerable populations.

There will also be changes to water supplies: while estimates as to the likely rise in sea levels vary (depending on the level of water expansion and whether or not the ice caps and glaciers melt), most experts believe some rise is likely — probably about 40 cm by 2100. While this doesn't sound like much to most of us, to those in low-lying coastal areas and flood plains it is potentially devastating. Many could face salination of their fresh water supplies and the inundation of fertile growing areas, such as the Nile Delta. Consider too, how many sewage treatment works and large industrial complexes are sited on the coast, as well as densely populated ports and resorts. Given that the water table under many cities is already rising, higher sea levels will only add to existing problems. Greens have always said that the London Docklands development is not sustainable — global warming could prove us right even more quickly!

All these changes are also likely to precipitate a significant movement of people. This will not just be in Africa or Asia, where we in the West can look upon it as their problem. It will be ours too. Consider how much of Europe is populated at the coast or on the banks of major rivers fed by mountains and glaciers. In Switzerland, the glaciers are already retreating. Look where we have situated Sellafield, Cap la Hague and Sizewell. We need to take global warming seriously.

The largest volume of gas contributing to the greenhouse effect is carbon dioxide: about two-thirds comes from burning fossil fuels, the remainder from burning wood. The situation is exacerbated by the loss of the great forests which absorb the gas and by changes in land use. Methane, another contributor, is produced through cattle and rice farming, the decay of certain wastes and emissions from certain industries. CFCs also contribute, absorbing radiation more readily even than carbon dioxide. Low-level ozone, created by the action of sunlight on aerial pollutants, mainly from car exhausts and industry, is another significant contributor.

Simply by looking at the origins of many of those gases which are now disturbing the balance of our climate, we can see how much our industrialised way of life has created the problem.

Politicians know that if they want to delay the onset of global warming in any way or perhaps lessen the more dramatic effects foretold by some (involving the total disappearance of some countries, such as the Maldives), they have to take significant action fast. The Rio Summit came up with the Framework Convention on Climate Change, which came into force in

March 1994, after the fiftieth country had ratified it. The Convention urged countries to stabilise emissions of carbon at 1990 levels by the year 2000, amongst other things. The 1990 levels were already problematic. Some countries will get nowhere near that modest target and some of those will be amongst the biggest emitters.

The general expert view is that to stabilise atmospheric concentrations of greenhouse gases will actually require a cut in their output of some 60-80% below current levels. As many of the world's poor and developing nations do not emit anything like the same proportion of greenhouse gases as the industrialised countries, it is obvious (well, it's obvious to Greens and some others) that those countries will have to cut back still further.

Economics in the real world

Such measures cut right to the heart of the energy use which drives our economies. According to some economists, they are likely to mean an end to economic growth as measured by GDP. It is such an enormously sensitive issue that the vast majority of political parties throughout the world do not want to tackle it seriously and they accuse the Greens of scaremongering when we want to talk about it. Who's 'living in the real world'?

Our manufacturing, trading and transport systems are built on cheap energy supplies, mainly from fossil fuels. It is a measure of political success to deliver cheap energy — you only have to look at what is demanded by and from OFGAS (the gas industry's regulator), where a £30 reduction in the average gas bill is seen as success for the consumer. We do not demand overall energy efficiency from the utilities or require that they reduce greenhouse gas emissions as part of their service, we only look at a narrow economic definition of success. Indeed, large consumers of energy are often charged at cheaper rates per unit, which can encourage overconsumption.

The Green movement in the 1970s was deeply worried about how long our fossil fuel supplies and other natural resources would last if we continued to use them so rapidly: would we be able to extract oil reserves in the Antarctic, say, and at what cost to the area's ecology? Greens then also became concerned with the problems of pollution, such as acid rain, and its effect on forests, lakes and wildlife. Global warming has become the next set of issues associated with our use of fossil fuels — and we have not really solved the problems related to unsustainable use and immediate pollution.

As a species, we have developed our ability to exhaust our natural environment. We have gradually become aware that we have done this in the past — that the Sahara desert was once the grain basket of the Roman Empire; that humans created the American dustbowl; that the extinction of the Dodo came about because we killed too many, and so on. What we have done in the last century is to increase the rate of exhaustion and the implications for the world as a whole are serious indeed.

I remember a number of years ago sharing a platform with Jonathon Porritt where he shocked the audience by saying that he felt single issue campaigns were of little use: that there was no point in campaigning to save the whale, if the next year you had to campaign to save the tuna fish, and the next to save the mackerel, and so on. What was needed was a campaign to change the system that made such campaigns necessary.

I was reminded of this when I saw the Greenpeace campaign early in 1996 to save the sand eel and other small marine life forms from being transformed into fish oil, which makes its way into so many products, including cakes and biscuits! These life forms are the basis of the marine food chain: when they are vacuumed up from the sea for our profligate consumption it means that other species die or are reduced in number because they have insufficient food — a marine famine. Because the system has not changed, more creatures are threatened. Politicians have not responded, so again the pressure groups are forced into action.

When we add this to the factory fishing of the world's seas and the overfishing of so many species, we have a disaster. Many coastal communities in Africa, Asia and South America rely upon fish as a staple source of protein in their diet and as a valuable part of a truly local economy. Their in-shore fishing grounds are virtually fished out in many parts of the world as the huge factory fleets of Japan and other countries move in. The EU has struck deals with many African governments to allow EU boats to fish in African waters to keep catches up and national fleets at work while at the same time reducing the catch quotas in European waters in pretty pathetic attempts to preserve viable fish stocks.

Fish farming is now becoming a growth industry, with all the knock-on effects of intensive farming in terms of pollution, the overuse of antibiotics and so on. But the development of fish farms can also distract attention from the underlying problems of over-fishing elsewhere and over-capacity in fishing fleets. These problems are further exacerbated by the lack of effective international agreements on regulating and policing fishing activity.

About a third of the fish taken from the seas are not even used directly for human consumption but are mainly turned into animal feed for pets, livestock and pond-raised fish. This export market is often more lucrative for poorer countries than their own domestic market, so a valuable food resource for local people is exported, usually along with the profits.

A water crisis

There is also a problem with the way we use fresh water. Modern development tends to be water intensive. We make liberal use of water in the home (for washing, flush-toilets, green lawns), in our leisure pursuits (for swimming pools, golf courses, sports pitches), in industry (in smelting works, cooling towers, fabric production, paper making and many other processes), and in agriculture (which still accounts for about 70% of water use).

The supply of fresh water is becoming a source of tension at the international level and within countries. Many of the aquifers upon which Israel relies for its agricultural production are in the West Bank area: Turkey has dammed the Euphrates and can effectively control much of Iraq's water supply. Even in the UK we have seen fresh water brought by tanker from Northumberland to Yorkshire to avoid the need for stand-pipes in the street.

In 1994, Japan's hottest summer on record, water shortages were so severe that many utilities and manufacturing firms around Tokyo were importing water from as far away as Alaska.

Countries such as Saudi Arabia are using their underground water supplies (sometimes called 'fossil water') to supply agricultural needs and leisure activities. Such water is very hard and slow to replenish. Even many of the wells dug to alleviate local poverty in parts of Africa and Asia are beginning to take their toll of the local water tables.

So we are using our supplies faster than we can replenish them. This is simply not sustainable. Wasting water has also become an issue: in the UK up to an estimated 30% of our clean water supply is lost through leaks. We are also polluting our fresh water supplies — whether through amounts of sewage too great for the system to handle; salination, as a by-product of poor irrigation schemes; or chemical pollutants from industrial or agricultural use.

In many ways we are also using our fresh water supplies inefficiently. This can range from what seems like very small-scale usage, such as the amount of water taken to flush a toilet, which becomes large-scale when multiplied by millions of toilet bowls. To use water of drinking quality for such a purpose is unnecessary and expensive; some (such as Prince Philip) would argue that we use too much water for such an exercise, while others would argue that flush toilets are not needed anyway, as there are alternative technologies.

At the other end of the spectrum is the biggest fresh water use: irrigation. Poor irrigation schemes damage the environment. By large-scale diversion of rivers in order to irrigate cropland, the flow of water into those rivers is affected with multiple effects on wildlife, the build up of pollution and silting and by the leaching of agricultural chemicals into the water. The Aral Sea, between Kazakhstan and Uzbekistan, is an often quoted example of the deleterious effects of poor planning. Here as a result of the diversion of its two feeder rivers to irrigate land for cotton, rice and vegetable production, the inflow of water was severely reduced; the surface area of the sea has reduced by over 40%, its volume has reduced by two thirds and salinity levels have trebled, and the 24 species of fish once commercially fished there are believed to be extinct. Winds pick up salt from the lake and annually dump an estimated 43 million tons of it on surrounding land, reducing fertility still further.

Much of the damage caused by irrigation was done through concentrating on how to get water onto crops, with too little thought given as to what the water would bring in and take away and how it would drain. Large loans were given for irrigation but, as no environmental impact assessments were required, only half an effective system was actually funded. As usual, the cost of rectifying mistakes is enormous — if indeed it can be done. Some agricultural land is now having to be taken out of production because it is too salty to grow crops. We have created wastelands.

The problem with agriculture

The desire for intensification of our farming processes is one of the root causes of such problems.

We have seen it in the gradual increase in soil erosion figures, whether for East Anglia, the West Coast of America or the Sahel. Such erosion is estimated to be undermining the productivity of one third of the world's cropland. That is a staggering figure, with awesome implications for our ability to feed ourselves.

What causes it? Partly the loss of organic matter holding the soil together; as artificial fertilisers have replaced dung, compost and other vegetable matter, so the topsoil has become more dustlike and therefore more easily dispersed. Partly the loss of trees and hedgerows, which act as windbreaks and contain the soil's dispersal. Also, a change to more intensive farming, which has required greater output from the same land, allowing less time for recovery or rotation — more cultivation breaking up the soil, ploughing more deeply and fragmenting its structure, and chemical rather than organic inputs. In many parts of the world, growing pressure for more agricultural land — whether to meet growing population needs, or the displacement of people from traditional farmland as it is brought into cashcropping, or goes under concrete — has meant a move into more marginal land, less suited to agriculture.

It can take from 100 to 250 years to create an inch of topsoil: we can disperse it in only 10 years. What a waste.

We are so anxious to squeeze more from nature, that we have failed to consider the consequences. An old teaching colleague of mine used to quote an Estonian proverb: 'You can only rob the land for so many years and then it will pay you back.'

Yet many are still looking in that direction as a solution to our problems. Greens in the European Parliament have been at the forefront of opposition to genetic engineering. They have successfully helped to stop the introduction of BST (Bovine Somatropin) into the EU. This is a hormone which encourages cows to produce even more milk, through even more distended udders, and wears them out even faster than our current dairying methods. This is called progress — to make each animal into a more productive mini-factory unit. The long-term effects of BST consumption in humans is not known.

We are seeing the production of pigs, genetically engineered to have eighteen teats to suckle their increased litters (pigs generally have between eight and twelve to a litter), so that each farrowing is more productive and profitable. We have seen the 'creation' of hybrid animals and are supposed to be impressed by such scientific genius.

Crops and animals are being engineered to be disease-resistant, yet many of the diseases are themselves a by-product of intensive rearing. We shall soon be able to buy tomatoes developed to last longer on supermarket shelves and better constructed to withstand long journeys without damaging their skins. Presumably the search is on to develop calves with a similar property — better able to withstand obscenely long journeys to the slaughterhouse.

Genetic engineering offers a shortcut to a myth — plenty for all at no extra cost. It may be that there are some advantages, but many people feel that there is inadequate understanding of the medium to long-term effects of such developments. How will the new varieties react in a wider ecological setting? What will they do to other existing species?

The rush for genetically engineered products is a reflection of the way we have shifted from agri-*culture* — where the use of land is part of a whole way of life and a natural cycle — to agri-*industry*, concerned with maximising production and fixing every problem.

It is no wonder that we are seeing a growing number of vegans and vegetarians as more and more people are revolted by the way in which the agricultural industry treats livestock in particular, demonstrating an almost total lack of respect for other species — exemplified in the feeding of sheep's carcasses to cows, a naturally herbivorous species, and the resulting epidemic of BSE.

Food at what cost?

The food production industry sees added value, not in producing better quality food through more sustainable agricultural practices, but in more highly processed food. The bulk of EU agricultural subsidies goes, not to farmers, but to food transporters, storers and processors. The CAP, even after the McSharry reforms which aimed to shift support to income from prices, is still wasteful. We have seen agricultural land taken out of production through set-aside, which has left a more intensive form of production in place and which has benefitted larger-scale farmers who can afford to have up to 10% of their land out of production (albeit compensated for) and still have a viable farm left.

When that is added to the inefficiency of crop use, in feeding cattle on purpose-grown crops when that land could be used to provide crops to feed people (roughly 40% of the world's grain is fed to livestock and there is a double land input — for the crops and the cattle) we can see that a diet high in animal fat and protein has high environmental costs. Those costs accrue through the water, energy and fertiliser inputs into growing and transporting the feed and meat. In some parts of the world (such as the Sahel and Brazil) there has been an increase in soil erosion and desertification, caused by introducing or increasing cattle production to unsustainable levels.

The transport and energy costs involved in food production have also lead to concern about 'food miles' — the distances travelled by your food from the field to your plate. We are literally widening the gap between producer and consumer and replacing it with a fractured relationship, where the retailer is the go-between and thus accrues enormous power in interpreting the needs of each party. Five retailers — the large supermarkets — supply over 50% of the food market in the UK; fast-food chains also have considerable power in the food business. While it can be argued that their standards have helped to raise the overall quality of food sold to most people, they also control the type of food sold. It is their needs which determine the varieties of vegetables produced — for their ability to withstand long-distance travel, their colour, their uniformity, their consistency, their price, their availability and eventually their taste. In some deals, producers whose goods may taste fine, but don't look right, may suddenly find that their agreement is a sale-or-return one. Supermarkets are proud of their premium meat brands, produced in a more environmentally-friendly way, but the bulk of their animal produce — including, it can be argued, dairy produce — is factory farmed. Many of the cheaper, 'shaped' products (most frequently fed to children and to those unable to afford more) are often not so attractive when their content is studied in depth.

The argument is that we want cheap food. It is a very different argument from the one you might hear elsewhere in the EU, where the emphasis might be on safety, environmental concerns, taste or quality. What we are failing to realise is that, when all aspects of our food production are taken into account, it is not as cheap or as efficient as we might hope.

The political choices made about what constitutes efficiency in food production should be of concern to us all as they affect our health and the health of our planet. It seems that politicians only worry about agriculture when scares affect our domestic and export markets or Euro-sceptics make a fuss about our payments to the CAP.

Political choices made internationally encourage the environmentally inefficient (such as monoculture, chemical dependency, large-scale developments like dams and huge road systems) and also contribute to the loss of biodiversity (the range of plant and animal life in existence).

In the UK we are contributing to the loss of biodiversity in our own country, just as the loss of tropical and temperate rainforests and coral reefs are doing in a more dramatic way elsewhere. We are only slowly coming to realise how vulnerable that loss may make us.

The crops on which humanity relies are based mainly on 30 kinds of plant out of about 80,000 potentially edible species. As we see them changing and developing through our intervention, so the threats to them from disease and new strains of insect infestation change: global warming and ozone depletion will add to those threats. Scientists have argued that we need a large gene pool to provide insurance through other options. We are being told that the possibility of finding new medicines for human ailments is helped enormously by ensuring wide biodiversity.

These are the pragmatic arguments for maintaining habitat but there are also ethical considerations. Do we have the right to knowingly destroy the existence of other species or even bring them to the brink of extinction?

Greens would argue that we do not. We are a part of nature, albeit the current dominant species. As a part, we too are susceptible to evolution and extinction, we have a responsibility to the future and, if you subscribe to the selfish gene theory, we want to be part of it. Green politics is about trying to find a way forward which will still give us choices in the future.

·7·
The new framework

We face enormous problems in many aspects of our lives, yet we feel our politicians are not getting to grips with them. There is no collective desire to really tackle environmental degradation and poverty, or to provide worthwhile work, and there is certainly no apparent sense of urgency.

Yet all around us we see the evidence of these problems, and global communication ensures that none of us can escape some knowledge of what is happening elsewhere in the world. There is a growing belief that politics is about more than income tax — it is also about our wider hopes and our vision for the future.

We want those who represent us to be in touch not just with our daily lives but with our deeper concerns and values, and to have these reflected in the solutions offered to us. We want to feel we are listened to and included in the process of making decisions about our future. We want to feel we matter.

So it is not surprising that so many people are turning to the Green movement for inspiration and for answers: not just in Europe, but increasingly elsewhere in the world too. Why is this?

I believe it is because Greens are offering a new sense of purpose, and new goals in politics which resonate in people's hopes — a more secure future based on meeting real needs, and ecologically sustainable.

Sustainability has several ideas within it — capable of lasting but also able to support and to nourish, so it represents a particular form of development. In its deepest sense, too, the nourishment is not just of the body, but of the spirit — a sense of completeness and meeting all our needs. So this is a key concept for Greens.

When the then Ecology Party published its core document back in 1975 and called it 'The Manifesto for a Sustainable Society', 'sustainable' was not a word with any widespread political use. Indeed, one local paper published an interview with an Ecology Party candidate at the time of the 1978 local elections in which the word 'sustainable' was replaced with the word 'suitable' throughout. If the editor thought 'sustainable' a difficult or non-existent word, what did the readers make of the concept of a 'suitable society'?

The document set out our ideas as to how we could achieve a way of life which respected the needs of people and the rest of nature, aiming to ensure a safe, secure and lasting future. We applied the scientific idea of ecology but gave it a political meaning — to show how policy areas were inter-linked, not free-standing, and that we had to integrate environmentally sound goals into our thinking and planning if we were to have a future worth living.

Green politics is not single-issue politics but an integrated, holistic politics concerned with all dimensions of life.

A sustainable society cannot be one based on injustice or inequality in which people lack the means to survive. Such societies are by definition not stable, carrying within them the basis of their destruction.

The definition of sustainable development most widely used is that set out in *Our Common Future* — the Brundtland Report for the U.N published in 1987: '...meeting the needs of the present without compromising the ability of future generations to meet their own needs.'

This is an important definition as it looks to concepts of global fairness, conservation, and the long-term. All are key elements in Green thinking.

However, as so often happens with language, many politicians have made a half-hearted effort to understand and ended up mangling the concept of 'sustainable development', or have taken a deliberate decision to bastardise it. 'Sustainable' is a warm, secure word in a world of uncertainties, so we now see a new political concept attached to an old one as in 'sustainable growth', frequently used by David Owen, now adopted by Kenneth Clarke, and we can all feel happier that nothing's really changed. And make no mistake about it: it hasn't.

There should be a real political battle over the meaning of 'sustainable'. Conventional economic growth is not sustainable, it consumes more than it produces, so it cannot last. Politics in future will increasingly be about the new political goal of ecologically sustainable development.

Meeting that goal will require a different range of perspectives in politics. Increasingly, we need to see the inter-linking of issues — a new political ecology: this is gradually being recognised and the environmental and women's movements have been powerful catalysts for this. Both the UK and the European Union's internal workings are now supposed to have cross-commentary and linking in policy areas on environmental considerations and a woman's perspective.

Another area of change, again one in which environmental concerns have played a major part, is that of international co-operation. Our recognition of global interdependence has been accelerated as we try to tackle the problems of acid rain, marine pollution, global warming and ozone depletion. Our international relationships are not a question of choice but of necessity. This is an uncomfortable fact of life for parties with a strong belief in the supremacy of the nation state as they struggle with the need to share and combine power in the interests of our common future.

This also means that the more far-sighted are coming to deal seriously with the issue of subsidiarity — placing power at the lowest effective level. This is not the same as 'small is always beautiful', which is a frequent mis-representation of Green politics. Greens are about *appropriate* scale and we prefer to look locally and regionally before we look nationally or internationally; but we are a global movement, and take seriously our responsibilities to the planet. To us, the local is a more human scale unit, where we can work together for our common future: if that basis for trust is under-used and neglected, the foundation of our common home is weakened.

True subsidiarity is about being prepared to use all levels of decision-making as appropriate to solving problems and ensuring the progress of ecologically sustainable development and to be clear as to why that level of decision-making is appropriate: it should not be a defensive or power-seeking mechanism but a keystone of good government, giving clarity and helping to engender trust.

If we are to aim for the new political goal of ecologically-sustainable-development, our thinking needs to take on a longer-term perspective. This is something not encouraged in a system which feels 'a week is a long time in politics' and where it is all too easy to become engrossed in many small pieces of legislation, without being clear as to their overall purpose, but maybe that's because our big political parties no longer know what their purpose is. Ecologically sustainable development provides a benchmark against which we can measure potential legislation: is it a positive step towards that goal or not? What might its effect be in five or ten years time, or even longer? What will be the likely effect for future generations?

A longer-term view encourages responsibility in politics and imposes a greater collective responsibility on politicians if they have to look at the possible outcomes of their decisions and justify them. A real sense of the future also tends to emphasise preventing the occurrence of problems or feeling you have a solution to them before you set something in motion: for example, if there had to be a solution to the problem of disposing of long-lived nuclear waste before we built the power stations, they might never have gone into production.

A new political goal, encompassing different perspectives, also requires new 'tools' to make it happen. These will include different indicators in order to assess a changed view of progress; new or reformed institutions to carry policy through; and a new legislative framework within which to operate.

A change of culture

To be totally effective, however, such political change has to be rooted in its wider culture — a culture which values co-operation and the experience of others, and where we see ourselves as part of the natural world, and feel a responsibility towards it. A culture in which both men and women can play an equal part, free from the stereotypical roles into which they are pushed.

The Green movement itself draws its inspiration from a variety of sources: from the Quaker tradition, from the influence of Gandhi and Schumacher, from many great religions and supporters of the Gaia philosophy, as well as pagan and humanist thinking. Members bring their experience from the feminist movement, from working overseas, from the inner cities and many other places. What binds us is a feeling of connection with the Earth and a desire to create a new relationship with it and all its inhabitants. It is a movement centred on the question 'how well?' rather than 'how much?' and it looks to co-operation as the way forward. Greens want to create a way of life in which there is time and space for silence, contemplation and the development of relationships with each other and the planet.

There are those who would argue that this is unrealistic — humans are basically selfish, competitive and chauvinistic. Maybe that makes such people feel comfortable about themselves and their own attitudes and disappointments but there are also plenty of signs to show that's not the sum total of human behaviour or aspiration.

So how can a different political perspective and culture help to solve the huge problems facing us?

·8·
Tackling poverty: a new economic direction

To tackle poverty effectively we have to recognise it as a real problem. When I listen to certain politicians these days, particularly in the Labour Party, it seems to me that poverty is a forgotten word and the poor a forgotten constituency that isn't talked about because it might frighten other voters or because their votes are taken for granted. Yet study after study, and our own common sense, prove that our failure to really grapple with deprivation now only faces us with enormous additional costs, particularly in terms of spending on health and social security and wastes the lives and potential of millions. It's an issue we must face domestically and in global terms as the gap between rich and poor widens.

We need to think hard about how to meet the needs of the human population within the context of ecological sustainability, living within our means, and this will have enormous implications for those of us in the richer parts of the world. Our energy-rich and environmentally wasteful lifestyles will have to change and our expectations along with them as we come to grips with new economic realities.

Our economic thinking will have to adapt to creating an economy which aims for the most efficient use of resources possible and much slower depletion rates: we will have to safeguard and maintain our renewable resources and minimise and account for the waste we produce as we aim for a *conserver* rather than a *consumer* economy.

This search for true sustainability will mean we need to emphasise subsidiarity in trade — importing nothing we can produce ourselves. This will mean a redirection of our economic thinking, which sees international trade as a driving force in our economy and the amount of it as a key indicator of the health of our economy. But emphasis on such trade can make us vulnerable, dependent on too many external forces, which is unwise in a time of growing ecological instability.

We need to develop stronger and more diverse local economies, here and elsewhere, not dependent on a single 'cash-crop' employer: we want local economies that can keep people and their cash in the community. This gives a more balanced economy nationally and thus reduces the pressure on other areas in terms of providing more jobs, more housing and more related services than they can sustain.

Strong local economies, however, are not an excuse for keeping nasty foreign goods out — a xenophobic protectionism (our global environment proves we cannot live in a vacuum) — but will provide a shift in our priorities towards protecting our environment and meeting our basic needs more effectively than at present. In fact, strong local economies are more equitable in global terms as they give greater economic self-determination to the world's poor countries.

There is obviously a clear need for government action in setting and achieving new goals. However powerful consumers and industry may be, they do not have the power to shift economies in a concerted way and deal with the effects of such shifts: nor do they carry the political legitimacy to make this happen.

But, in order to develop policies which really solve problems, governments must have a clear picture of what's really happening in society and what the trends are. As I've already explained in Chapter 3, the key indicators used give a misleading picture confusing quantities with quality. We need different indicators in order to develop a different economy.

One such measure is the Index of Sustainable Economic Welfare (ISEW), pioneered in the USA by economist Herman Daly and theologian John Cobb. It assesses whether economic performance over time has delivered a better quality of life. Like GNP, it starts from a measure of personal consumption but then refines this in a number of ways: it weights the figures according to income distribution (a pound to the poor is more significant than to the rich); takes account of 'defensive expenditure' — spending to offset environmental damage (noise pollution, air pollution, ozone depletion, cost of car accidents, etc.) — and includes estimates of the longer term cost estimates of environmental damage and resource depletion. It also includes a value for household labour.

Including the 'invisible work' done in maintaining and caring for the household is in line with numerous UN resolutions over the years to recognise this work, predominantly done by women, on which our economy depends. Few countries have yet carried out the necessary changes in their accounting.

When the ISEW is applied to the UK (as done by Jackson and Marks for the Stockholm Institute and the New Economics Foundation), it becomes clear that, although we may have more money, our basic quality of life has improved very little. Why not? Because we are spending more to offset environmental damage and to maintain basic welfare levels, as well as because there are some things that personal expenditure can't buy — such as clean air — which significantly affect our quality of life.

The United Nations Development Programme is also beginning to employ a different indicator — the Human Development Index. This includes factors such as life expectancy, literacy and purchasing power rather than GNP. It also looks at what the standard of living really is for people and indicates whether they have the necessary skills and power to change it. The UN's *Human Development Report* of 1996 highlights a growing disparity between rich and poor and points to greater instability as the likely outcome.

Thus, the use of different economic indicators, giving a clearer picture of people's daily lives, reinforces the Green perspective. If we want an ecologically sustainable economy, we have to tackle poverty and the widening gap between rich and poor.

But as well as measuring differently, we must price differently. If we are to encourage the sustainable, we must discourage the unsustainable. This means pricing goods and resources to take account of the environmental impact they have throughout the production process and after it. Instead of externalising pollution and disposal costs (leaving them for someone else to pay), those costs must be included in the price to the company and the customer. How this could be done will be examined in more detail in Chapter 11.

While the British government is being comparatively slow to introduce measures to enact the overtly environmental aspects of a greener economy, it's not doing too much to improve the human factors of a sustainable economy either. Despite record levels of spending on social welfare, the poor are getting poorer. In the formal money economy, there are few benefits in kind and these are largely for children — milk tokens, free school dinners (although these are no longer available for children whose families are on Family Credit and authorities no longer have a duty to provide them), school clothing vouchers, etc. The emphasis is on providing cash to highlight the individual's responsibility to spend wisely and to give the individual some choice over spending. This may be fine in theory but is not much use if your purchasing power isn't enough to cover the basics. Benefit payments don't cover the unexpected like repair bills or replacing stolen or damaged property.

So, given that we expect people to meet their own needs in a cash economy, how can we ensure basic security?

In the short term, while we're using the current mix of benefits, we need to:

i) ensure people can find out their entitlements: restore the freephone advice line, increase the accessible advice sessions (such as those held in libraries, pubs, community centres), provide clear literature.

ii) provide well-trained, well-paid and sufficient staff to deal promptly with claims, giving accurate and consistent advice.

iii) integrate the current benefit/tax/national insurance systems: no one should be worse off if they do paid work, study or volunteer regularly. We need a sliding scale to cover NHS charges (if they remain) so that earning marginally more means you don't pay them in full.

iv) increase Child Benefit by 50% and retain it for 16-18 year olds in education.

v) re-introduce benefit entitlements for young people; remove the differential in Housing Benefit for under-25s and allow students to claim benefits.

vi) increase basic benefit levels to raise purchasing power and to compensate for higher prices caused by environmental tax measures.

vii) change income tax levels: increase the basic tax allowance by 50% and have three levels of income tax payment: standard, intermediate and top.

The guarantee of a basic income

In the longer term we should implement a Basic Income Scheme (BIS) — a universal basic benefit with additional income taxed progressively as proposed above, but removing the need for tax allowances and benefits.

The BIS gives a degree of independence to every individual and is a form of state recognition of work currently unpaid. It goes to people in their own right and is not based on marital status or work history. It recognises the fact that income, not work, is what meets basic needs in our form of society, even if taxation on work and resources provides the overall pot of cash.

Many things are claimed for BIS: that it gives a means of support to those in an abusive relationship and therefore a chance to leave; that artists, performers and athletes can concentrate on their work without being forced to take up other work; that people can take study breaks if they choose or work on projects of benefit to the environment or community. Perhaps most importantly, it helps to blur the lines between the status accorded to those in paid work and those who are not.

Others claim that it is a low-wage charter, allowing employers to pay less than a fair wage. Well, the current system already does that — in a clandestine manner that forces people to break the rules and risk losing benefit or staying poor. A basic income would at least allow openness about additional earnings and give people more choice about whether or not they take low-paid work.

A minimum wage level for certain work would still be an option but loses its attraction if pay differentials are maintained and prices subsequently rise. It also deals exclusively with those in work. Many of the poor — children, elderly people, those unable to work through some form of disability — are still poor. The issue is about meeting needs and purchasing power — not the amount of money as a bald indicator.

BIS also addresses some of the tensions in our current system. Single parents (mothers, mostly) are seen by some as a drain on the state who should be at work; yet parents should care for their children and not be at work when their children are at home. Everyone gets BIS, so they have more choice. Of course, decent child care facilities would also enable people to make a real choice: tax breaks for workplace crèches, well staffed playschemes and after school provision would also help meet the needs of children and their carers. Such measures would also enable child care to be shared more equally between men and women.

We are told that work is good for you. Poor people, we are told, should get less benefit so they have an incentive to do work, even if it is low-paid; but rich people, the theory goes, will only work if they pay low taxes and are therefore allowed to keep more of their money! BIS says we are all inherently worth the same respect and should all be taxed progressively.

There are questions surrounding the level of BIS and how it might be paid for but there is increasing interest in the idea (it is now also Liberal Democrat policy) because it does do a great deal to relieve the poverty trap (whereby people are worse off by working) and to remove the current grey area of entitlement, where people can slip through the net and find themselves in deep financial trouble or destitute — a particular problem for so many young people at present. Also, because BIS is an entitlement rather than a benefit and not means tested, it should not be seen as charity or something shameful, a view which currently stops some people claiming benefits.

The LETS alternative

Pounds sterling are not the only form of income which can help meet basic needs. Increasingly, Local Exchange and Trading Schemes (LETS) are being introduced at local level. These are used as a way to bring those without standard income back into the local economy using a nominal local currency to pay for goods and services. People setting up a LETS decide on the unit of currency (Strouds in Stroud, Bobbins in Manchester, etc.), its nominal value and the amount of local currency a member should start with. They set up a register of members' skills, services and tariffs: members are then free to trade amongst themselves and pay accordingly — in only the local currency or a mixture of that and the national currency. Others may join as the scheme develops.

LETS are growing rapidly. Whereas they started as informal (i.e. outside the formal economic structures) and small-scale initiatives, that is now beginning to change: local councils such as Manchester are becoming interested in getting involved and the DSS and Inland Revenue are becoming interested in the paid work and additional income aspects.

For many people, LETS are providing a means of meeting a variety of needs: basic repairs can become affordable if you don't have to pay for labour in standard currency out of your low income; an allotment can be maintained providing fresh food; child-care can be provided at times; new skills can be learned; equipment hired out; hair cut, etc. Payment in local currency frees more standard currency to meet bills, etc. It is a means of maintaining your skills base during periods of unemployment or developing additional skills and interests which may enhance your work opportunities in the formal sector. LETS can provide a reasonable income for work done at home — unlike much home work which may pay only pence per hour.

LETS are also a way of integrating people into an economic community and making them feel they can participate in and benefit society in some way. Thus the schemes are important in developing self-esteem and social cohesion, the lack of which is costly in personal and economic terms. So, while not a complete answer to meeting all needs, LETS have a contribution to make and are a means of developing local economies.

Keeping finance local

As more banks and building societies pull out of the High Street in rural, inner city and more peripheral urban areas, and the government clamps down on DSS payments for basics such as cookers and beds, people are looking for ways to provide financial services at the local level.

Credit unions, a nineteenth century initiative, are growing in popularity — there are now over 600 in the UK. They currently have assets of some £65 million and loans amounting to £50 million. They are a form of 'save and borrow' cooperative which can deliver small-scale loans (which larger institutions often find it difficult to cope with) at comparatively low rates of interest. They are not interested in persuading people to take out loans in order to increase bank profits, but are there to lend on what people really need. Some are trying out debt redemption schemes where they will 'buy' a member's debt and repay it, the member then repaying the union at a nominal interest rate. The potential benefit to hard-pressed local authorities in not having to pick up the pieces after, say, the bailiffs have been in, persuaded South Glamorgan County Council to provide an initial £3000 for such a scheme in Ely, Cardiff. The default rate is very low, despite the fact that they are often lending to poor people, which demonstrates that being poor does not mean you are irresponsible and feckless.

Community banking is also becoming more popular. Initiatives in the US, Ireland, India and elsewhere have shown that people are prepared to save with local banks that lend to local businesses. Like credit unions, the amounts involved need not be large nor the rates of return particularly high. People are keen to see a sense of vitality and purpose in areas which may appear run-down and hopeless. Such initiatives keep cash in the community and people can see the benefits of their efforts: they can also be run in a more accountable way by involving local people in their policy making.

Local food coops are also finding a place — bulk-buying a range of produce, including fresh foods, and selling on at lower prices so that people can eat more, better or the same with a little more cash for other necessities such as water, heat or housing.

There has been a growth in the number of Black and Asian business associations willing to advise and support each other and provide training opportunities for those who have all too often found it difficult to get financial and other support for their business enterprises.

Local government can also help to foster the local economy through the provision of premises for small and medium scale enterprises, and by providing local economic development units to offer free or cheap advice and training to businesses. But the range of possibilities is often stifled by the wrong sort of regulation.

Compulsory Competitive Tendering (CCT) is an example: local authorities should have the right to place contracts with local suppliers and businesses if they wish to support local jobs and enterprise, and have the right to stipulate that such contracts are dependent upon certain conditions — such as equal opportunity practices or getting supplies from environmentally sustainable sources. The cheapest is not necessarily the best choice. But the Conservatives have forbidden local authorities from applying these criteria when placing contracts.

Work for all who want it

Since most people in industrialised economies gain their livelihood through paid work rather than by being personally self-sufficient, employment is a key factor in meeting needs. Greens believe in *sharing* the work that needs to be done amongst the people who wish to do it. If all such people are to have access to paid work, that means changing our attitudes and expectations about work in a positive way.

For most of us, work is a means to an end: we may enjoy what we do and derive great satisfaction from it but most of us have other aspects of our lives that can give us that. Increasingly, we are seeing trade unions (such as ASLEF and the RMT) bargaining for shorter working weeks. There is an EU Directive (fought tooth and nail by the British government on our behalf) limiting the working week throughout the EU, as work becomes more complex and stressful and unemployment grows.

Greens do not subscribe to the 'poor law' work ethic of the Conservatives and, increasingly, the Labour Party — that you are only entitled to a share in society if you work for it in terms defined by the government — the 'work-fare' philosophy. Such an attitude fails to recognise all the strands which make up a healthy society and which can't be measured in the same way.

Greens are not interested in creating full-time, paid work just to keep us occupied and improve the national growth figures, particularly when so much of that work encourages consumption for its own sake. Greens believe in doing work that needs to be done — food production, housing provision, environmental protection, etc. — to meet our needs and to function constructively as a society. We believe it should be 'good work' that has a positive purpose and maintains our self-respect. A Green work ethic centres on doing necessary work for the benefit of us all, that may or may not be paid for in the standard way; it is a contribution to society, not a condition of entry to it.

So, how do we share paid work in a sustainable economy?

Firstly, by removing a number of barriers to employment itself, encouraging more flexible employment and finding areas where employment can grow to improve the quality of our lives — recognising that in a Green economy, job opportunities will shrink or even disappear in the more environmentally destructive industries.

Greens advocate the abolition of the employer's National Insurance contribution and the integration of the employee's contribution into income tax. There is no point in taxing employment when we want people to have work. We should be shifting taxation to activities we want to discourage, such as waste, pollution and the unsustainable exploitation of resources.

We support the moves to put part- and full-time working on an equal footing in terms of employment rights and pay: part-time work is still valid work, not a second-class (and usually female) substitute. Job-sharing is also a viable way of working in many circumstances, although many employers are still reluctant to take it seriously. A more flexible attitude to the retirement age would also make sense, allowing people to move to part-time working before retiring completely: it would also help to retain experience and help to pass it on.

Employers and trades unions also need to look at jobs in which they automatically expect overtime to be worked and for that to be necessary to provide a living wage. Regular, 'built-in' overtime means you should really have more staff and a better wage structure (initially helped by redistributing the employer's NI contributions perhaps); recommended limits on working hours are often linked to health and safety considerations and shouldn't be taken lightly. We should also look to eliminating 'zero-hour' contracts where we can.

In terms of increasing the available paid work, we need to look at how we can intensify the human element of such work — a necessity in a more sustainable economy which aims to reduce fossil energy use and the input and waste of natural resources.

People often talk about the 'four Rs' of a Green economy: refuse, re-use, repair and recycle. There are a lot of opportunities in the last three of these. Refusing unwanted and unnecessary products and services helps set the climate for the other three.

Re-use of an item does not have to be in your own household or work environment but can be done through it being handed or sold on. A vast number of charity shops have appeared which provide voluntary work and training opportunities, paid work for key staff and a cheap source of products — helping us meet certain needs (for clothing, furniture, household goods) for less. There are a number of charities and businesses equipped through the second-hand market for computers and office furniture. The only bottle re-use scheme we have is in the dairy industry, unlike countries such as Denmark where the presumption is in favour of bottle re-use.

The main job opportunities come in repair and recycling. If we want to extend the useful life of many things, an efficient and effective repair sector is essential. Repair possibilities also have to be made standard and designed into new products as part of a more sustainable lifestyle (producing goods to last, not to discard).

But repair and maintenance of our building stock and infrastructure is also of importance. Schemes such as the 'Care and Repair' projects (maintaining the fabric of the homes owned by elderly people so they can remain there) have a long-term future and require a variety of skills.

Road maintenance is currently under-funded yet even in a future which sees more walking, cycling and public transport, there is a need for good quality pavements and highways to ensure comfort and reduce wear and tear.

Recycling is an area of the economy which is still under-developed in the UK. Government targets have been set for local authorities to reach in terms of domestic waste (50% of recyclable waste to be so treated by the year 2000) but markets are patchy and volatile at times. Until we have a taxation system which charges virgin products at their rate of environmental impact, recycling will not reach its full potential. The landfill tax, which came into force in 1996, is a very small step in the right direction. Countries such as Germany have already introduced pricing mechanisms to make recycling more economically viable. Here, because the market is comparatively small, those that want to develop schemes such as recycling agricultural plastics have been forced to transport such plastics miles to the single recycling plant able to deal with them. While the government has supported some pilot recycling projects, yet there is general resistance to tackling the issue of pricing — an essential part of changing the economic framework.

Yet recycling can be comparatively labour intensive, again offering jobs at a variety of skill levels from collection and sorting to the re-processing itself. According to figures quoted by the Worldwatch Institute in their *State of the World Report, 1995*, one million tons of solid waste can produce 1600 jobs in recycling, 600 in landfill and a mere 80 if incinerated.

Changes in energy use can also create jobs. We need to use energy more efficiently which means much better insulation programmes. The Home Energy Conservation Act 1995, drawn up by the Green Party and ACE (the Association for the Conservation of Energy), introduced into parliament by Plaid Cymru's Cynog Dafis and taken up by the Liberal Democrats, in particular Alan Beith and Diana Maddocks, requires local authorities to draw up energy audits and set efficiency targets for their areas, aiming to reduce wasted money and resources. The initial focus of remedial work will be the most energy-wasteful homes and the fuel poor, who spend a disproportionate amount of their income on heating. Such programmes will also help to reduce the harmful emissions from burning fossil fuels.

Meeting such targets creates jobs — again at a variety of skill levels —

in both producing the materials needed and carrying out the work, and it helps to improve the lives of the poor so that they are more comfortable for their money or can release some money for other needs.

There are also jobs in energy production in a Green economy. The emphasis in future will be on renewable energy sources some of which can work efficiently on a much smaller scale than the current power station system. Solar panels can be installed on individual buildings; combined heat-and-power schemes can provide energy for neighbourhoods; wind and wave power can also be operated on a variety of scales. There is research, development, construction, fitting and maintenance work to be done for all of this — again offering work at a variety of skill levels. Practical experience elsewhere in the world and numerous research studies demonstrate that smaller-scale and renewable energy supply initiatives are more labour intensive and, being localised, keep more money locally.

As we aim to reduce our energy and environmental costs, public transport becomes more important. There will be more jobs in this sector too in construction, maintenance and staffing, offering alternative employment to many of those currently employed in the car industry.

A less energy intensive, more organic form of agriculture will also require more labour, helping to keep workers in rural areas where the loss of work has been rapid. More people employed in agriculture can also mean more job opportunities in servicing their needs locally.

Greens take maintenance of our infrastructure seriously. We are possibly the only political movement genuinely interested in sewage — its treatment and potential use. There is a need for a massive overhaul of our, often Victorian, system. How we supply our water is also of primary interest to us and there is room for investment in delivering it more efficiently (reducing leaks, the possibility of more ring-mains and storage facilities etc.) and in introducing technology into our homes which will allow us to use some water twice before it goes into the waste-water system. So there are construction and maintenance opportunities in our infrastructure.

Fair trade

A shift to a more sustainable economy, in terms of meeting more of our needs locally, will also reduce our demands on other parts of the world, particularly the poorer countries, allowing greater self-determination there.

To help establish greater equity, the rich world has to relieve the debt burden of the poorest countries. Where loan agreements have been entered into by governments representative of the people, it would be more constructive to have repayments made in *local* currency into a *local* development fund, than to force countries to export in order to earn foreign currency. This would free countries from struggling to meet IMF re-structuring criteria and instead allow them to invest in primary health care, including family planning facilities, locally efficient agriculture and education programmes. There is significant evidence that investment in social and personal development raises living standards.

A further step towards a more realistic economy is to pay fair prices for products and resources, reflecting adequate wages. A Green economy demands that we have to take a realistic view of the overall cost of things if we are to make the correct decisions for the future. So we should pay fair prices for (for example) copper, gold, tea and coffee that also enable the producers to pay their work force a living wage and to minimise environmental damage. There have been advances in the Fair Trade movement, often led by development charities, and helped by large-scale organisations such as the European Parliament buying in fairly traded coffee and some supermarket chains stocking CaféDirect. However, there is still enormous scope for expansion of Fair Trade and it has to become the dominant ethos, even if such trade is limited to things we cannot produce ourselves.

Hand in hand with fair trade must go policies to control the large-scale Trans National Corporations (TNCs). These wield enormous financial and political power: you only have to look at the history of the European Single Market to realise that it is a concept devised and promoted by TNCs to enhance their market opportunities. They need to be controlled because they effectively constitute a second, unelected government and, it could be argued, often constrict primary governments by giving financial support to particular campaigns or candidates, promising investment if certain policies are followed or threatening to pull out if they aren't; because they can buy up a country's natural resource and exploit it with little real investment in that country and then walk away from the consequences; because they can employ variable ethics according to different regulatory structures, not always applying the highest standards across the board; and because they control about 70% of world trade and are therefore critical to the achievement of fair trade and ecological sustainability.

Economies are distorted because profit doesn't stay where activity produces it. Greens therefore argue for the repatriation of profits, so that communities that suffer the consequences of production also benefit from it. Governments should be able to institute capital controls to ensure profits stay put. This will have even greater significance now that the GATT covers Trade-Related Intellectual Property (TRIP) allowing companies to own what was previously common property such as genes and species. We could see a cure for AIDS being developed from an African source, patented by a Western company with no benefit returning to the country of origin. While the aim must be common ownership of intellectual property from such sources, in the short term we must provide benefits to the people of the country of origin.

Governments should be co-operating to insist that the highest regulation standards become the norm for any TNC operation. This reduces the possibilities of social or environmental dumping of either location or product — so there would be no export of chemicals banned in the EU to poorer countries, for example. Such international agreements would also work against variable health and safety standards in, say, the oil industry.

The longer term goal has to be the break-up of huge TNCs in order to counteract their near monopoly power in some areas and their stranglehold on pricing. This could be done through an international Monopolies and Mergers body. There is no doubt that we need the re-introduction of the UN's Centre on Transnational Corporations to monitor their activities and to ensure the enforcement of international agreements.

Keeping the controls local

Looking to an ever more unstable future, you can understand why the governments in the EU want to introduce a common currency — it has a feel of solidity, reassurance and power and helps cement the regional trading bloc more firmly. Greens argue, however, that we need to retain financial flexibility if we are to move towards a more sustainable future. We reject a single currency as removing that choice. We can see arguments for a parallel currency — the Euro running alongside national currencies — and being the appropriate currency for international trade. Given the Green preference for more localised markets, a more centralised currency is not what we want.

We are sympathetic to the idea of taxing currency transactions over a certain level and paying that revenue into the national purse or the Global Environmental Protection Facility. Governments should have the right to use currency controls to protect their local or national economy as an aid to greater self-reliance.

Greens also want the flexibility to vary rates of VAT in order to encourage sustainable activities and discourage the less sustainable. We want to see a zero rate on housing repairs and energy efficiency measures, for example. That cannot be done under the current EU agreements, which aim to harmonise VAT levels as part of the Single Market project.

However, as moves towards the international economy continue, financial decisions are removed further away from the people whose lives they directly affect. This is just one of the reasons why Greens want financial decisions to be controlled at an open and accountable level. It is why Greens are helping the work for different and reformed international institutions which will be looked at in more detail in Chapter 10.

One of the financial developments which is helping to bring about change is the growth of the ethical investment sector. This, at last, allows people to bank their cash in a way which does not support companies with poor track records in environmental protection or human rights. For those with money to invest, it can be used to support greener enterprises. Other investors, such as the large pension funds, have begun to question the regulations which require them to invest money in the most profitable way rather than ethically as their members might wish.

Some lending institutions now offer differential rates of interest to investors so that they can invest in interesting but more less profitable enterprises at a lower rate of return, thus providing a wider range of financing possibilities. Some organisations, looking for investment, have sought it through conventional shareholder channels like the Centre for Alternative Technology at Machynlleth or through enrolling supporters as co-operative members like the Out of this World chain of ethical supermarkets. Wider forms of ownership, such as co-operatives or community businesses, provide other ways of attracting investment and need support to expand. In a greener economy, the main providers of investment capital are likely to be community banks rather than the large commercial banks as we now know them.

So the ecologically-sustainable economy will be doing less but fairer international trade which takes account of the environmental and social costs.

It will emphasise the more local economy and aim to develop jobs which are sustainable and useful and to share this work.

It will recognise the value of the unpaid and informal economy and aim to provide a basic income to help all members of society provide for their basic needs.

It may seem a long way from meeting needs to changing global trade agreements but the framework in which we operate has to change. We have to remove the barriers which stop us doing what we need to do in order to solve our problems. A cultural shift in our economic thinking is needed if we are to start from what people need, not what the international economy currently demands.

· 9 ·
Caring for communities

The UN Convention on the Rights of the Child, which Greens feel should be part of our domestic law, includes:

• the right to learn to be a useful member of society and to develop individual abilities; and

• the right to be brought up in a spirit of peace and universal fellowship.

As so often with Bills of Rights, the language is aspirational, a reflection of how we wish ourselves to be. How can we achieve such goals for our children and for ourselves?

Education for life

We are expected to develop our individual abilities through education and then demonstrate these abilities in our work. That gives a narrow definition of 'abilities', as if they are only academic, but we are endowed with creativity, the capacity to develop relationships and to affect the wider world through our choices.

Does even our current education system develop our abilities and help us learn to be useful members of society? I don't believe it does — restricted as it increasingly is to making us theoretically fit for conventional paid employment.

Firstly, education is something that needs to continue throughout our lives in a variety of ways. The University of the Third Age and the growth in the number of mature students demonstrate that people have a thirst for knowledge throughout their lives; different experiences mean we develop new interests or make us want knowledge or understanding that seemed irrelevant when we were younger. A need or desire to change career may also lead us to more education. We need therefore to plan this into our political thinking. Local skills exchanges can do this on an informal level and adult education classes have long been a fairly cheap and valuable way into studying. There are more schools now willing to open their classes to adults and many more could follow, as could universities. The growth in independent studies and the continuing enthusiasm for the Open University are ways in which educational structures are opening up for the better, making formal schooling less of a 'one-chance saloon'. Mature students are now in the majority on some university courses.

Training, however, is more of a mess. Part of the problem is the short-term contract culture and the decline of manufacturing industry, with not much stepping in effectively to take the place of long-term on-the-job training — as formal apprenticeships were. Good vocational training requires depth and time, not a 'patching-up' approach: it needs status and a variety of levels so that people can fully develop their abilities.

As more jobs are created in sustainable areas and we are looking to increase labour input, reduce mechanisation where labour is more efficient and develop the repair side of work, it should be possible to re-develop the apprenticeship model — albeit in a modernised form. We need to develop the theoretical framework of some vocational courses so that people can adapt their knowledge to new situations, and we need to re-develop craft skills training.

In terms of examining ability in particular areas, these demonstrations of skill and understanding will be set as appropriate and be graded in difficulty. There will still be a need for external validation and a general understanding within the trade or profession of what is involved.

Within schools, there is less need for a public examination system in the old sense. As further education becomes more diverse, both in content and means of entry, there is no single convincing argument to justify the domination of the school curriculum by the requirements of universities. It is a pernicious system which leads to an undervaluing of some skills and areas of study at the expense of others, and encourages the belief in a hierarchy of intelligence based on exam results. The absurdity of the current system is obvious every summer when we have the ritual 'it's all too easy' call of the blue- and red-feathered nostalgia birds. Exams should be about attainment: if you've reached the standard, you are entitled to the qualification and pass marks should not be adjusted to ensure that a set percentage of people fail. Let's go for levels of achievement instead of GCSEs. These are already in use in many lower school (ages 11-14) courses and are built into the National Curriculum (of which I am no great fan). If you can accomplish certain tasks, demonstrate certain skills, etc., then you are on that level — no arguments, you deserve it. It's a clear scheme, providing incentives, and says what you can do — which exams (particularly without course work) don't really make clear.

Those achievement levels shouldn't stop at whatever age you leave school but could be added to later and when you're ready, breaking the ritual of collective hysteria in the summer. It also allows for a more flexible school-leaving age. There's no virtue in keeping people in buildings and chasing truants to force them in there just to make it look as if they're learning. Over the years a number of schemes have been tried which eased pupils into work or college as they outgrew school: for a number of students they proved valuable and it is time to reintroduce them. We need to be far more imaginative about education: yes, there are basics which people need to learn but they're not all given weight in the current system and we have a habit of assuming that a couple of hours a week over five years is the best way to organise all learning — but where's the evidence for this?

People need to learn to read, write and work with numbers — basic skills that bring fulfilment and open so many doors. We need to learn how to work with other people and resolve conflicts. In 1978, at the first UN Special Session on Disarmament, Britain and 178 other countries agreed to the inclusion of peace studies in school curricula; we have yet to meet that commitment. Personal relationships are important too: our schools need to provide high quality sex and emotional education and to prepare children for parenting. Such education needs to give value to both homo- and heterosexual sexuality.

We need to learn about the world around us, its environment and other inhabitants — part of being 'brought up in a spirit of peace and universal fellowship'. We should learn about food production and nutrition so that we can understand the effect of what we eat on our environment and ourselves. We should also learn how how to prepare food (a subject being edged out at present). And let's prepare it *together*: virtually every society in the world places importance on communal eating as a way of bringing people together.

We must understand about the technologies that now shape our lives, not just how to use them but how to control them and understand the wider implications of their use.

We also need to learn how society works and how we can participate as useful members. We can practise these skills in school councils, in groups that help to deal with internal conflict and help to devise school policy: schools can form links with the community and use the skills 'out there' to help educate their pupils.

If we're to be educated for life, we should be developing interests that may be life-long: sport and fitness activities, creative activities, scientific enquiry, gardening, mechanical skills, language learning and so on.

Every young person should leave school with a sense of self-worth and self-respect, understanding the effects of their actions and able to take responsibility for them.

Schools can be very exciting places, but to really come alive they need to be liberated, not confined to league tables and form filling. There is a strong argument for a basic education, to show a range of what's possible, but beyond that, and in conjunction with further and higher education and training bodies, schools should be more concerned with 'developing abilities'.

There will be some parents who will not wish their children to attend school and some children for whom it doesn't work and who may therefore be educated at home. Those parents should be assisted and local children's centres can be used to provide places for children to socialise and offer assistance in areas of learning where parents feel less confident. Such centres are also helpful as places for support and advice for parents and children alike.

Schools should be accessible to every child in a local community, support being provided to help all children benefit to the best of their ability. Segregating physically impaired children does not help develop a 'spirit of universal fellowship'. But if that's deemed by child, parent and school to be in the child's best interests then there will be alternative facilities available, but the presumption will be for education in the local school, unless an 'opt-out' is requested. That's the only kind of educational opt-out Greens are interested in.

As a party we don't mind private schools if they're set up on educational grounds (Steiner schools, dance schools, etc.), rather than to grant social privileges. I don't believe schools should be run as charities. Nor do I believe state funding should go to schools who push a particular religious line: by all means let's learn about different religions and other belief systems (humanism, for example) in order to foster common understanding; let's set aside quiet space in schools for meditation or prayer and let us discuss and develop the more spiritual side of our nature. However, for me, religious indoctrination has no place in schools and nor does single-sex education. The long term benefits of single sex education are far from clear, particularly for boys, and schools should encompass as full a community as possible.

So a sound education system throughout life that helps people for work, leisure, personal and social relationships and that aims to develop individual abilities is an important factor in building self-reliance and self-esteem. Concentrating on the positive contribution people can make opens possibilities and helps include people rather than exclude them. How can we build on that?

The role of a written constitution

Whether we succeed or fail will depend on the framework within which our society operates. This basically decides overall priorities, sets the relationship between the different levels of government and gives the formal means of access to this system. It is more important than many people believe.

We need a modern, written constitution that sets out the relationship between the people and government so that it is clear to all, not just those who operate the system. That constitution must make it clear that power resides in the people, not the Crown. There is no justification for the hereditary principle allowing preferential access to the levers of power. If we want to retain a monarch in order to meet important guests then we should decide that by referendum, as any key constitutional issue should eventually be determined. We may decide we want a directly elected president with limited powers (possibly OK) or lots of powers (and lots of battles with parliament? — no thanks) or to rotate the job in parliament or round the regions — rather as the EU presidency rotates among member states. A second chamber in parliament (having primarily a revising and constitutional role) would be directly elected.

Our new constitution should set out the relationship between different levels of government, their areas of decision making and their revenue raising powers: there's no point in making decisions if you can't pay for them.

This would be putting subsidiarity into practice and mean a lot more decentralisation, with power at the local level where decision-making should be rooted. We want to see assemblies for Scotland and Wales as steps to total independence if their people so choose, and that possibility in England too in the longer term. We have always felt that regional government should evolve across the UK (and that means within the constituent nations too), preferably based on bio-regions (each including a watershed, agricultural land and varied natural resources) as a move towards greater self-reliance. The national government becomes the co-ordinator of energy production and transportation, foreign and defence policy, management of bio-diversity (including animal rights policy), natural resources management and human rights policy — amongst other possible co-ordination roles. Social services (in the widest sense) become the role of regional and local government and so on.

Obviously, election to all levels of government would be by proportional representation (PR), so that the people's views would be fairly represented. PR also tends to elect a higher proportion of women and, because it offers a greater variety of candidates, also gives greater possibilities of representation to people from ethnic minorities, physically impaired people, or any other minority group currently under-represented in parliament. The voting age should be lowered to 16 (if you can pay tax and consent to sex at that age, you ought to be considered responsible enough to vote too) and we should introduce a rolling electoral register that includes homeless people, registered through the town hall itself if necessary, so that as many people as possible are able to exercise their right to vote. PR would make it more worthwhile.

We must remove the financial deposit required to stand as a candidate in parliamentary and European elections: let those who want votes persuade people to nominate them if they think they should be on the ballot paper. Democracy is partly about the relationship between people and their elected representatives: it shouldn't be about the size of your wallet — so let's put a national limit on the amount parties can spend trying to get elected and keep paid political advertising off our TV channels.

Another item on the constitutional list must be a commitment to a real Freedom of Information Act, not subverted by claims of commercial confidentiality or bogus national interest. A democracy cannot function without information, which really is a form of power. Decisions about the safety of chemicals and drugs, defence spending or food safety cannot be made if some groups have the information and others don't. You cannot be a truly useful member of society if you do not have the possibility to make informed decisions.

A Bill of Rights is also an essential element of a constitution. That now has to include a responsibility not to abuse our environment. There is considerable debate about what should be included in a Bill of Rights and indeed whether it should exist at all — it would give 'too much power to the judges' it is said. A reform of the judiciary, giving more ways into it, a more open selection procedure and a duty to make it more representative of society at large, would probably help people have more faith in it and not just in terms of a Bill of Rights. If we also give the directly elected second chamber a duty to assess proposed legislation against a Bill of Rights, that would also remove some of the onus from judges.

Much debate centres on whether to include 'social and economic' rights, such as the right to a job or adequate housing. However, many feel that this gives too much political power to the interpreters and can lock you into a system of organisation which may no longer be appropriate, particularly in the light of changes necessary to reach an ecologically sustainable society. Environmental rights are also controversial — not just in terms of the rights to conditions necessary for survival such as clean water, adequate nutrition and shelter, but because they raise the question of whether the environment has rights as such.

I believe it is necessary to have the 'survival' sort of environmental rights in a Bill of Rights. As the gaps in society widen, we see water disconnections and evidence of malnutrition, both of which could be challenged under a Bill of Rights. An ecologically uncertain future means we cannot take such things for granted. I have yet to be fully convinced on 'Rights for the Environment': if we were serious about access to clean water and adequate nutrition, we would already be seeing a much Greener society. However, we need to find a constructive framework for expressing our responsibility to the planet.

Given, too, that the areas least likely to participate in the current democratic framework are those of high material deprivation, it can be demonstrated that there is a link between the political and survival rights. As it was put to me once: 'If you're spending your time scratching a living, you're too tired for anything else.'

The civil and political rights, over which supporters of a Bill tend to agree, cover areas such as the right to fair treatment under the law, the right to vote, demonstrate and associate (as in trades unions and political parties) and to freedom from discrimination on a wide variety of grounds — Greens would include sexual orientation, language, economic status, age and having a travelling lifestyle, in addition to the more usual categories. We have also argued for a general anti-discrimination clause in the EU treaties as an aid to anti-racist and anti-semitic work in particular.

Thus a Bill of Rights gives each citizen a clear focus and means for challenging the power of the state, whatever their social status.

To this constitutional framework, we would also add the right of citizens to call for a referendum on a government decision if a certain number of signatures were gathered to demand it and similarly, if a certain signature threshold were reached on an issue of concern to the public, the government would have a duty to conduct a referendum. The Citizens' Initiative, as this is known, is already in use in Italy and elsewhere: it has been used to force referendums on the electoral system, pesticide use, nuclear power and hunting, amongst other things. So it can be a tool for the people to make politicians discuss what they would often rather ignore.

A written constitution also addresses nationality or citizenship. Greens want to move to the point where we have a citizenship based on residency criteria, rather than nationality based on an historic blood relationship to a country.

Asylum rights should also be enshrined in a written constitution. We have a responsibility to carry out in both their letter and spirit the international agreements we have signed up to. The recent changes to our asylum law are shoddy and mean-spirited. We must have a fair and speedy resolution of applications with a genuinely independent appeals procedure and a right to those benefits available to citizens (including the right to work) while the process is underway: detention should be an exceptional state of affairs. Let's end the Carrier's Liability concept (using air crews as immigration officials on the cheap). We should be working to extend the UN definition of a refugee to encompass that used by the OAU which includes those fleeing occupation or gross civil unrest.

Those who want to savagely restrict the number of asylum seekers argue that we're a small island — we are, but do they honestly believe the 23 million currently living outside their home countries, the vast majority in Africa, want to come here? There is a carrying capacity argument (how many people can we support on the resources we have?) but until we are really living in a way that gets us off the backs of the poor elsewhere in the world, I believe we should honour the obligations we've signed.

As for the argument that 'more refugees will worsen race relations', there's an interesting assumption that refugees are black — which is what the argument is often really about. Well, it doesn't do much for race relations to see more black people living on the streets, to know they're in hiding or to know they could be deported to a country where they are likely to be at risk.

I believe I have shown that a written constitution is of genuine importance in making us equal before the law, with clear rights in respect of our security and our right to participate.

But a constitution has to live on the ground if it is to mean anything. It can reflect or help to create cultural change but that only comes when people feel it in their bones. *How can we make it real?*

Changing the culture

Increasingly, people want to give their time to improve their society. The government has made many changes in the organisation of basic services that require people to give their time if those services are to function. So, we can value the voluntary work that keeps our society going. Government can ensure that the right to time off, attendance allowances, dependent care and training is written into law so that the formal parts of our system, relying on volunteers, can function: work such as being a school governor, being on a Community Health Council or housing association board of management.

Those, however, who serve on government or locally appointed quangos, should be appointed by an independent selection panel on the basis of publicly available criteria.

The voluntary sector is huge, from youth groups to Age Concern, and involves millions of people: yet training staff and volunteers to serve within it can be expensive. We should be looking at all levels of government for support and more grant aid for projects. Giving local government more control over revenue is an essential part of Green thinking: one of the reasons is to enable local communities to support their own initiatives.

There's so much energy and goodwill in our communities that groups should have the time and energy to do the work — not spend so much of their time fund-raising — that's a real employment growth area at present. Voluntary groups could be funded on a longer cycle (say four years) and should be able to handle their own cash directly.

Government could do away with the pre-condition of matching funding — if the job needs doing, then fund it.

Planning for people

If we want neighbourhoods to feel like communities, we must re-think the design, from the casual meeting points — the post boxes, the bus stops — to the more formal areas such as local shops, playgrounds and sports fields. We need to create public space that is welcoming in terms of its accessibility, design and use. Such spaces can become threatening when they are anonymous — miles from where people live — or the vast piazzas in front of the towering office blocks which are ghost towns after work. We should be designing out dead space.

That's why the re-integration of our communities is important: bringing together our work, leisure and living activities so they interconnect and faces become familiar. This means changes in planning, with less artificial zoning. There is still much more use to be made of school buildings for sports facilities and meeting rooms (schools not having to charge large amounts would help) and many places don't have the community centres they need. We still have too many pubs, and too few cafe-bars which cater for a wider age group and are not so alcohol-oriented. Reducing alcohol intake among the young, especially in public places, would also make the streets safer places and reduce crime.

Young people often complain, quite rightly, that there is little to do. We need to listen to their needs and work with them to provide solutions: it should be seen as cost-effective to provide well-resourced youth facilities rather than do nothing and then complain when they turn to anti-social activities. Many councils have had to cut their youth work budgets due to government capping which to Greens represents a suppression of local choice and should never be allowed.

Local theatres are often unused; we see comparatively few local art exhibitions or musical events. Yet there are many people with creative talent for whom there is little local outlet. Clubs and dance venues are often restricted to over-18s because they are licensed. Evening classes often exclude the school-aged and young, and the elderly might not want to be out after dark, yet 'twilight' or weekend activities are rare.

We don't find enough meeting points, particularly between generations and between cultures.

Access for all

Our transport priorities, too, need to change, as we're gradually coming to realise. We need to shift from mobility for some to access for all. But when I hear everything blamed on the car, I get cross. Yes, there are too many travelling too much but lorries are still a huge problem, with axle weights due to rise soon. We need to get more goods off our roads and onto rail (but that means better and more road/rail interchanges); we could also put more onto the waterways. The EU is currently looking at both these areas. However, while people have been out protesting effectively and with great commitment against road building, I have yet to see an imaginative demonstration for a goods yard. Overall, however, we need to reduce the need for so many long road journeys, which means the reduction in long-distance trade and different pricing policies mentioned elsewhere.

We need to invest in better, more regular, clean, safe and cheap public transport — accessible to all, from wheelchair users to parents with small children. This will primarily mean buses which are more flexible than other multi-passenger carriers. There's certainly a place for rapid rail and tram links (people tend to like them and are therefore willing to use them) but there's also a place for cars — as taxis and, for some jobs and in some parts of the country, as the most efficient form of transport.

People will reduce the use of their cars when it becomes more expensive — the trip to the local shop or school will then be done on foot; when employers stop insisting on car use for certain jobs which could equally well be done by bike, bus or taxi; when cycling isn't so life-threatening (fewer cars and lorries, more cycle lanes) and when walking feels safer (fewer bikes on the pavements, fewer cars and more familiar faces) or when public transport becomes more attractive. Also when the shops, workplaces and leisure centres are closer to home. Some of these changes require national government action, some can be brought in locally. The joint aims have to be to provide alternatives and to reduce the overall need for journeys.

In some places, 'walk-to-school' schemes are being introduced where a system of guardians (parents, householders en route, more crossing patrols, etc.) keep an eye on children's safety along the way. It's imaginative schemes like this which help people feel more secure about leaving their cars and which develop the local community. Traffic-calming, too makes parents feel happier about encouraging their children to walk.

Coping with crime

'Guardian' schemes have come into being in order to help reduce crime. Neighbourhood Watch is the most obvious; some places introduce 'car watch' schemes while meetings are going on (too dangerous to come out on foot). Better street lighting and more police on the beat are the usual solutions. Street crime has sapped the confidence of many, who are reluctant to go out or — if they're younger — may carry weapons. There are no easy solutions at all.

There are citizens action schemes in the States where neighbours, often parents, get together and reclaim their buildings — by standing firm, refusing access, clearing up, calling the police, and making sure vacant flats are occupied: with sufficient mass and energy it can work.

Some of the successful neighbourhood approaches come from a nucleus who are often too angry to take it any more, or who are worried for their own children — they are not willing to be passive. They are often women, who have not had their formal status and worth defined for years by paid work but by their ability to hold things together and be creative on a small scale. Hope comes through increasing the range and power of small-scale democracy, the power to affect your own life in your own locality. The small scale is where men and women must both now learn to work as the formal working world changes. Groups working at this level, appropriately supported, can introduce gardens, playschemes, repair schemes, small workshops, education classes, food co-ops and community shops and run credit unions and LETS schemes and community centres. They make the community feel alive and as if what people do matters: reducing anonymity is constructive. All of this enhances hope and self-respect, and reduces crime.

Another factor which has to be worked through in order to reduce crime, is trust in the police and legal system. If local troublemakers are caught, what happens to them, how soon will they be back on the streets and what happens to those who give evidence? Fear of speaking up promotes a cycle of silence.

Part of the overall problem is a justice system which cautions or imprisons but where little restorative action is taken. Community service is seen by many, sometimes even the politicians who put it in place, as a soft option, and prison is increasingly only about punishment. Greens believe we should do things differently.

People should be in prison as a last resort and prison should aim to change their behaviour for the better: prisoners should be treated with dignity and our prisons should be adequately staffed to do a constructive job. Juveniles should not be in prisons, but in secure units run by social services. Wherever possible, we should be looking at restorative justice. This means supporting the victims throughout, so they are not a double victim both of the crime and of the criminal justice process which can marginalise them and fail to support them. And, if they're willing, the victims can be brought into contact with the perpetrators so that the latter can fully understand the effects of their action. 'Restoration' should come in cash, or some form of work or community service designed to make reparation to the victim. Constructive work also needs to be done with the perpetrator to change their attitude and prospects: offenders often lack literacy and numeracy skills. The Citizenship Foundation has been carrying out some interesting work with offenders based on developing a sense of civic responsibility and an understanding of the consequences of their actions.

We need to find communal ways to deal with much crime. As a result of communities, police, social workers, educators and local councillors coming together, positive actions can be taken before problems arise, Solutions can also be found: community work for their own community; prisons or detention centres close by so that links are kept with family and friends; discussion about suitable penalties, of help to the area. This is not to support vigilante justice but people's concerted action in conjunction with the formal authorities. Developing trust in this way also gives people confidence in the police and vice versa. Given that most perpetrators are caught as a result of information received, greater public confidence could result in more criminals being caught — the greatest deterrent to crime. The maintenance of confidence is essential if we wish to continue with an unarmed police force.

Reducing the number of arrestable offences would also concentrate resources where they're needed. No one should be imprisoned for the non-payment of fines — community service is more appropriate than this modern equivalent of a Victorian debtors prison. Small-scale possession of drugs for personal use should be a treatable, rather than an arrestable offence. The real targets should be the importers and dealers, yet we're cutting Customs and Excise staffing and have few people working on money-laundering financial fraud.

But part of the context within which crime is committed has to do with our consumer society. The more we acquire status through what we possess, the greater the divisions within society and the less opportunity some people have to develop their own future, the more difficult it will be to deal with crime. Our society increasingly encourages us to think in the short term and take what we can, while we can. We are living through a period where the message is 'we have to look out for ourselves', creating work, securing a pension, insuring our mortgage in case we lose our job. Until this culture begins to change and people feel more confident of the future and their place in it, the problems will continue.

Health care: free and preventive

A feeling of 'universal fellowship' would ensure we care adequately for the vulnerable in society — the sick, the elderly, those with psychiatric illnesses or addiction problems, amongst others. We truly need to develop real community care involving friends, neighbours and voluntary groups as well as the formal providers. We need to sort out the anomalies between social service and NHS funding of particular services and to establish a clear management and review process for people cared for in this way, in which the client also has a say.

There's a need for more staffing, and continuity of staffing in this area, preferably attached to fully developed local Community Health Centres providing a wide range of services.

Such care demands a high level of liaison between different authorities to avoid people being lost in the system and not getting the necessary care.

Care should be free at the point of use. While the aim should be to enable people to live in their own homes for as long as possible, the move to other accommodation (sheltered housing, nursing home) is not usually a totally free choice but one necessitated by diminishing faculties and the need for greater care — a health reason and, it could be argued, one based on preventative action to avoid potential injury or worsening health.

We must provide sufficient respite care and support services to enable carers to continue their valuable work, without suffering themselves. Crisis care and long-term care for the mentally ill also need to be improved, so that the vulnerable can be protected.

Greens want to see the emphasis in health care much more on illness prevention and health promotion. New indicators of social welfare will include health factors. This means taking environmental measures:

improved air quality will lead to fewer respiratory problems, lower speed limits and fewer motorised vehicle journeys will reduce accidents, while less pollution and better quality food will have a beneficial effect on health. We would also implement social measures to improve housing quality and income levels, together with a ban on all tobacco and alcohol advertisements to help reduce demand.

These proposals, taken in conjunction with improved health, nutrition and sex education, better occupational health measures, more sports facilities and an emphasis on walking and cycling in transport policy, mean we will see changes for the better.

Greens would encourage (and regulate) complementary medicine as part of the range of techniques available in a health service where the emphasis is on low-tech healing as much as the high-tech hospitalised side. Complementary techniques could make a significant reduction in the amount of drugs prescribed, particularly barbiturates. We have no time for those who measure health provision by numbers of hospital beds, and the current political emphasis on that is unhelpful. We believe that GP fundholding and internal competition based on cost rather than quality have not led to overall improvements in health care: the 'gagging' of some in NHS Trusts also means problems are hidden, not solved. The Trust Boards should be replaced by bodies comprising representatives of local authorities, users, managers and the workforce, reflecting the stakeholder principle.

Decisions about the priority and availability of care should be made openly. There are already examples of how this might be done from the USA where local communities have drawn up priorities after careful consultation and these are reviewed on a regular basis.

A better housing stock

A contributory factor to health and well-being is the availability of adequate housing. Enormous pressure has been put on the South and South East and other areas by our distorted internal economy; stronger local economies will help towards a more regionally balanced distribution of the population, and devolution of government (giving less power to Westminster) will also help relieve the pressure on the capital.

Our first aim should be to maintain and improve the housing stock we have, rather than looking to new build as a first choice. In siting new build, we should aim to use infill rather than agricultural land, in both town and countryside. We need a greater variety of housing from hostels, foyers (housing with employment assistance for young people), sheltered and supported units to homes for single people and larger dwellings for extended families. There is still a need for low rent social housing in rural and urban areas: the extension of the right-to-buy to Housing Association properties is a retrograde step which will further reduce social housing stocks. Councils should be able to provide this if they wish and support should be available for self-build schemes.

We also need to improve the quality of housing by introducing higher insulation levels and better energy efficiency measures: these are already criteria for a loan from the Ecology Building Society and other lending bodies should follow suit. A home energy efficiency reading should be available to new buyers or tenants. We can make much better use of passive solar energy, use more heat-efficient building materials, and install dual water systems to reduce our overall water use. Visitors from Scandinavia often find British homes cold in winter because they are so poorly insulated. At least the Home Energy Conservation Act (see page 66) will help improve the situation.

But it is not just the quality of our homes but the quality of our overall environment that affects our lives. Charter 88 has been carrying out a Citizens Enquiry which has been asking a wide variety of organisations what changes are needed in the way we are governed to help us improve our lives and have more control over what happens to us. It is likely that a strong role for local government will feature prominently.

Local responsibility and Local Agenda 21

Agenda 21 (a lengthy document arising from the Rio Summit) is concerned with the changes necessary in order to develop an ecologically sustainable society into the next century. It places enormous responsibility on local authorities to deliver many of these changes as a result of real dialogue with local people.

The UK is one of the countries to have put effort into getting Local Agenda 21 (LA21) off the ground. The response from local authorities has been variable but a considerable number have embraced it actively. At its best, it brings local people together to look at what needs to be done to bring about positive change — and that change can be very wide ranging as issues interlink. All sectors of the community should be involved or it won't work properly. By coming to common solutions, people feel they 'own' them and have a greater incentive to see them carried through. It is a common educational process leading to a better shared understanding within the community. I believe it is a process whereby people also come to see the barriers to progress — the limits to what can be done locally — and so they look for different approaches from government.

With more power locally, more could be done. Local government needs to be free to raise and spend local finance. Greens would like to see the current set-up virtually turned upside down with most revenue collected locally and some passed up. The introduction of land value taxation as a key source of local income would give more control locally and help carry through local planning. Land would be seen as a community asset and taxed according to its value, based on its use. The tax can be varied to encourage or discourage particular uses; land left idle with a view to speculative gain could be heavily taxed, for example. Organic farmland or a children's playground could be taxed at a lower rate.

Through the stakeholder principle (which Greens have long endorsed), there is greater opportunity for a more coherent interlinked policy approach: companies, amenity authorities and social service providers could no longer act in isolation or such secrecy if there were real public involvement and accountability.

Even without legislation, local people or councils can introduce public 'round tables' of different sectors to discuss issues (such as transport, youth facilities, care of the mentally impaired) and come up with a range of proposals. Councils or organisations can develop real multi-agency working to improve action on, say, racial harassment or a reduction in teenage pregnancies. Citizens' juries can be introduced, a number of people randomly selected from a pool, to investigate particular issues and provide recommendations. There have been a number of pilot projects already. Citizens' initiatives can be introduced without need for government approval.

If people are to feel that they can act as useful members of society, they have to believe that they have a value in that society and that they can have an effect upon it: they must feel that their abilities are being developed and used. If we are to bring our children up in a 'spirit of peace and universal fellowship', that spirit must be rooted in our local communities so that we feel free and confident to look outwards.

· 10 ·

Global insecurity

When looking for security on a global scale, just what are we trying to keep safe and how should we aim to do it?

Historically, nations have sought to secure their boundaries from invasion in order to protect their people and natural resources. Obviously, individual security is sought after still: people want peace and stability in order to develop. We want to be able to look to the future with confidence: we want something better for our children. We don't have that confidence if we are afraid.

However, while some still fear invasion, for many more there are other threats. Threats of violence from within from terrorist attacks on a bus, in a park, at the shops; attacks by your own government or political opponents, as in Iraq or Algeria; and the longer term threats to your life and that of your children through deprivation or so-called 'natural' disaster.

There is no easy remedy to many of these threats to our safety. We can aim to reduce the rationale for violence and refuse to support it: we can work for the rule of Human Rights and support democracy, the freedom to organise politically and non-violently, and strive for more open and tolerant societies. We can aim to reduce the materialist demands of so many industrialised countries and work for greater equity.

This may still not be enough for some who either fear the loss of their power or refuse to accept the free and democratic will of the people expressed through the ballot box. Governments and political organisations have a duty to accept the latter — even if they don't like the result.

A co-operative agenda

The problems our planet now faces need international co-operation if they are to be solved. Such co-operation means adopting different criteria and a different political ethos to that which has put the needs of a government's nation state before those of others: it requires less self-interest and more shared interest. This may prove problematic to countries who are re-discovering a national identity and self-expression after being subsumed into the Soviet Union and Eastern bloc and who are also discovering that free-market economies are tough. Before being able to consume freely, they are being asked to restrain certain forms of production and development. The world's poorer countries may also find this problematic. They will be looking for clear signals that rich countries will lead by example and mean what they're saying.

However, there are signs that many people throughout the world are increasingly comfortable in having 'multiple identities': for example, East Ender, English, European, even 'global citizen'. We might feel we have different loyalty levels depending on circumstances. Fostering that feeling of many intersecting identities will be increasingly important as we look at our shared concerns, which cannot be totally solved at only one level.

A desire for a peaceful future and safe environment draws us all together and can provide a focus for common action, being a powerful driving force for change. Opinion polls consistently show that people believe environmental protection to be an appropriate issue for international co-operation and institutions. It will be important to retain that feeling of public consent and involvement through clear information and perceived progress. Local Agenda 21 has an important part to play in this, in helping issues connect.

The goal of ecological sustainability requires a different rationale to that of international trade. In trade, we agree in order to compete in a ruthless activity where we hope we won't lose. Ecological sustainability requires co-operation so that we can all benefit.

So what are the rules for a co-operative future? How are we to achieve the peace and security we want?

Underlying any action has to be an ethical, consistent approach: that our overall aim is not to increase misery but to prevent it; to act in line with international agreements and our own parliamentary rules (so no more refusing to admit we've changed guidelines on weapons sales or not followed our rules on aid projects); and to honour the agreements we make and take action we know is right, without making it a condition that every nation must do the same thing at the same time.

This may sound facile and obvious, yet when you look at agreements made on technology transfers, aid to the Eastern bloc, carbon taxation, etc., you come across a whole sorry catalogue of words and only partial action in a whole string of countries. We might already be living in a rather better world if governments kept their promises.

The most obvious steps to improving security are to reduce and remove weapons and minimise the tensions likely to lead to war.

We could start with landmines — easily laid but removed with enormous difficulty. Four out of five people killed or injured are civilians. There is a growing movement to stop their manufacture and export, especially 'long-life' ones, which the British government has slowly and reluctantly come to support. To reclaim land for displaced people to be able to return to is essential, but the cost is astronomical when the area is mined.

Implementing the arms agreements we already have is essential. START (Strategic Arms Reduction Treaties) I and II would mean considerable decommissioning and destruction of missiles, silos, bombers, submarines, warheads and fissile material. While progress is being made, it is likely that Russia will need extra cash and other support to make it happen: the Ukraine and Kazakhstan have also requested assistance. Metals are likely to be recycled ('swords into ploughshares'?). The US has plans to use the fissile material in its domestic nuclear power plants, but the storage costs alone are likely to amount to nearly $3 billion over ten years!

Russia is also having to find cash to pay for its part in implementing the Conventional Forces in Europe Treaty which came into force in 1992. The US has also to pay out considerable amounts on this, as does Germany which has responsibility for the weapons left in East Germany.

The Chemical Weapons Convention came into force in 1995; it outlaws the possession of chemical weapons and requires the destruction of existing stocks. The cost of destruction is an estimated *ten times* the cost of production. However, the body due to oversee and verify the process and check out possible future production — the Organisation for the Prohibition of Chemical Weapons — is likely to get only one third of its requested staff and half its budget.

There are a number of lessons to be drawn from this experience. Weapons systems really cost more than the publicly admitted price — green 'whole-cost' accounting should be applied to make this obvious. The 'peace dividend' can look less attractive if decommissioning costs are included in it: those costs should be included in the defence budget and protected within it, so that cuts are made in other areas if the overall budget then seems too big. If we are to avoid the costs and technical and political problems of decommissioning and destruction in future (assuming that weapons haven't actually been used), it makes sense to ensure a tough verification regime to engender confidence in such processes. Better still, of course, not to continue to manufacture the things in the first place.

A further step which would also reduce the circulation of weapons would be to cut back severely on arms sales. As one of the four main exporters, Britain could lead. It should start by closing the Defence Export Services Association and removing the credit guarantees it gives to the arms trade (more than to any other export trade). Eighty per cent of British arms are exported to Third World countries, many as part of aid/trade agreements, as the Pergau Dam scandal clearly showed. Such agreements should be stopped: they have nothing to do with real sustainable development.

The first arms markets to be dropped should be repressive governments where we know the hardware is likely to be used to repress internal opposition rather than combat external aggression. We should also be refusing to export likely instruments of torture to such régimes.

Arms conversion and peace building

Alongside the reduction in arms production, there needs to be a serious programme of industrial conversion and a programme related to the closure of bases too. Governments have a duty to follow through the logic of their policies. The USA has such a programme (if not on a grand scale) and the ex-Soviet Union had one on paper which has fallen apart due to lack of direction and resources. France has been forced to consider a partial programme in the light of the decision to end conscription and the public concern at what will happen to local economies dependent on bases and on arms factories. The UK has no such programme. We need a minister responsible for controlling and phasing out arms exports and developing conversion plans, possibly through a 'round table' approach of producers, trade unions, aid agencies, the military and ordinary citizens. Limited funds are available from the EU through the KONVER programme which draws on social and regional structural funds. This programme has developed from a proposal by the Green Group in the European Parliament in the early 1990s for a more extensive programme entitled RECARM. It would be worth expanding KONVER and also offering such assistance to Central and Eastern Europe through the Europe-wide Organisation for Security and Co-operation in Europe (OSCE) before too many export markets are developed there.

Proposals for an international Demilitarisation Fund have been put to the UN Development Programme by Nobel Peace Prize laureate Oscar Arias. Such a fund would have three main aims — restitution of damage done, conversion of the arms industry, and peace building — and could offer assistance to countries lacking resources. It could be funded through a percentage of defence spending and/or cuts in defence budgets resulting from a peace dividend.

There is no doubt that sufficient resources put into restitution of war damage, in the widest sense of restoring or developing stability through building civil society and ensuring greater self-reliance, would be money well spent.

The UK should also play its part in disarmament and should either cancel the Trident project or, if it proves too late for that, ensure it is never fully commissioned; we should not be adding to the numbers of nuclear weapons. In fact, we should do the opposite. There is no military or moral justification for the UK being a nuclear power. If the deterrence theory of Mutually Assured Destruction ('MAD') were followed through logically, all countries would have nuclear weapons. That is dangerous, expensive and does nothing to enhance ecological sustainability. The UK should implement a policy of unilateral nuclear disarmament. While some might argue that lays us open to charges of admitting we are no longer a world power, well, that is the truth. It could be argued that it is a clear signal by an industrialised country that we need a different form of security in future and that nuclear weapons are not the way forward. It also combats the arguments of some nations that are developing nuclear weapons that Western powers are only interested in maintaining the old imperialist status quo. We should not then scurry for shelter under the nuclear umbrella of the US and NATO or an EU umbrella held up by France. We could allow a full inspection of our nuclear facilities, making it clear at last that there is no military connection.

Defensive defence

Renouncing nuclear weapons, reducing arms exports and converting substantial parts of our arms industry are important phases in changing the UK's defence thinking. We should be moving to truly defensive defence — reducing our capability to undertake offensive (aggressive) actions and deploying forces in such a way as to clearly signal our defensive stance.

We are more interested in 'in-depth' defence, looking at successive layers of resistance. Our long-term aim is non-violent citizens' defence rooted in strategies of non-cooperation and possibly subversion. This draws on the long-established concept of the citizens' army — each citizen having a part to play in the country's defence — and combines it with the Gandhian philosophy of non-violence adopted by many civil rights activists. It is a low-cost, lower technology option and forms the end goal of a phased scaling-down in military defence terms.

However, we recognise that there are many steps on the way to that. One is to place our defence thinking in a Europe-wide context, not in a divisive framework which sees Europe in two parts. That is one reason why we see the OSCE as the appropriate institution for developing defence thinking. It is a body already looking at non-military conflict avoidance and resolution. Greens believe that the UK should withdraw from the Western European Union — rather than helping to strengthen it — and from NATO.

We part company with some European Green Parties who want to strengthen the common defence aspect of the EU and then move it towards the OSCE — we prefer to go directly there. We feel it is easier to maintain a non-aligned stance in the OSCE (which is important for neutral countries who feel under pressure in the EU) and it sets Franco-German co-operation in a different context. Working through the OSCE would help to reduce the distrust between the EU and former COMECON countries of East and Central Europe. A number of these countries are already anxious to join NATO because they would feel less threatened by this traditional enemy if they were 'in' and more protected from Russia: they are concerned at the special status being accorded to their old master and feel that their new national identity and autonomy are still under threat. The OSCE is a 'flatter', newer organisation clearly committed (like the Council of Europe) to human rights, but also to conflict resolution: it does not have the same militaristic perspective as NATO, nor the same free trade perspective as the EU.

The OSCE is currently an under-valued and under-resourced body — we feel this should change. A regional body (in terms of a global not national region) has a valuable role to play in identifying and dealing with potential points of conflict and this is likely to become increasingly important. The Organisation of African Unity is trying to develop a peace-keeping role as part of its activities and others are likely to develop.

However, if these initiatives are not to become simply another aspect of the machinations of potential power blocs, new collective super-states, there is still a need for an over-arching body in the United Nations. There needs to be a final arbiter and an organisation which can develop and share best practice. As Greens, we would be prepared to see UK forces as part of a permanent UN peace-keeping force and rapid response unit. But we feel that a greater number of skills are needed than the ability to carry weapons. The military are already valued for their engineering and telecommunications skills. Medical help is welcomed but we feel there is a greater role for a judicial presence (judges, lawyers, etc.) in order to help the development of trust in a 'justice system': it is an idea we have long felt would be useful in Northern Ireland.

The role of aid

For people to feel confident about demilitarisation, they also have to believe in a reduction of threat The common goal of ecological sustainability helps to give that confidence. Through greater self-reliance, less damaging environmental practices, less competition and different attitudes towards consumption, we reduce the competition for resources. Different indicators of well-being (see Chapter 8) demonstrate changed priorities and the search for basic security. The UN Development Programme defines these as clean water, adequate sanitation, food security, basic education, primary health care, family planning and nutrition programmes. *All* aspects of this programme are important: there is an opportunity to limit the number of pregnancies and births if children are healthier and live longer, and increased literacy helps to underpin all these changes, but there is no point in vaccinating children if they then starve to death. The UNDP estimates that currently only 7% of bilateral aid and 16% of multilateral aid funds are allocated to such programmes.

Our government should have supported the so-called 20:20 Compact on Human Development put to the UN's World Summit on Social Development in Copenhagen in 1996. The Compact proposed that donors should target 20% of their aid and recipient countries 20% of their domestic resources to such human development priorities.

Most governments have yet to meet the UN aid target of 0.7% of GNP set down years ago; the UK currently manages 0.31% and still does better than many other countries! The Green Parties of the EU have regularly called for a 2% target for the EU, at least half of which should be devoted to projects benefitting or controlled by women and all of which should involve local communities. The Greens prefer working through and with Non-Government Organisations (NGOs) rather than governments who can be part of the problem — creaming off cash, diverting their own resources to weapons procurement as aid comes in or delaying help to areas which are politically unsympathetic to them.

More and better-targeted aid, however, is not the only way of improving security: debt relief and changes in trade policy can also help. Freeing land for crops for domestic consumption rather than for luxury exports helps maintain local markets and can also help reduce the flood of migration to the cities and the resulting health and other social problems there. It also means people are not pushed in the other direction — to marginal land unable to support them. Access to land can also help the establishment or maintenance of family practices which improve land quality rather than degrading it.

However, such changes in economics and trade require changes in the financial institutions set up and financed by governments to regulate trade and to support governments experiencing difficulty: the goal for these — the World Trade Organisation (WTO), International Monetary Fund (IMF) and World Bank — must be to further ecologically sustainable development and only governments can agree to that change of purpose.

It is crucial that our large intergovernmental institutions work in partnership for the same ends. At present, they are likely to find themselves in conflict — in particular, the WTO and the UN. The WTO has its own legal status and set of rules deriving from GATT, without any clause committing it to environmental or social responsibility. If it operates as the GATT Tribunal did — and there's no reason to suppose otherwise — environmental and social concerns will always be the losers to free trade. The WTO can probably undermine any other international agreement. This conflict of power has to be addressed and the WTO either transformed or dismantled.

All these financial institutions need greater transparency in decision-making and votes of equal value. The size of their purse should not determine the power of their vote.

There has been some improvement in the World Bank since NGOs really became interested in its working but there is still a need for environmental and social audits of its projects and their effects in the light of international agreements before full funding is granted. The bank needs to find ways to invest in small-scale projects initiated by local people. At least it now has an Information Centre and has created an independent inspection panel where private citizens will have the right to raise complaints about projects supported by the bank. It will be interesting to see if this really makes any difference.

The IMF should be concerned with stabilising a country's domestic economy and ensuring it meets sustainable development standards and so should be lending with a view to resource conservation, greater self-sufficiency in food and energy, and to develop import substitution.

Our government should be reporting regularly to parliament on our decision making in all these institutions. Given that they act in our name, we should know more about what is being done and holding the government accountable for these decisions. Indeed, our representative to the IMF should be nominated by the Overseas Development Ministry.

Reforming the UN

There is also, of course, a need for reform within the UN and a clarification of its purpose. Its charter does not mention the environment although much of its work is now based on environmental protection and the relationship between that and human activity: nor is population mentioned, yet again the UN has realised the implication of the issue as a crucial factor in the quality and sustainability of development.

Making ecological sustainability the focus of the UN's work would shifting the traditional balance of power within the UN and give a sense of common purpose. That should also lead to an ending of the old nuclear privileges such as permanent membership of the Security Council. A world moving towards greater security, including disarmament, should also accord status to non-nuclear powers. A more regional structure within the UN could help design a permanent regional balance into its Commissions and Councils. There should certainly be no great power accorded to nations who don't pay their dues or conform to Resolutions.

There is a need for a global body developing frameworks and standards in the critically important areas of the environment, agricultural practices and health; in helping to create a more peaceful world and to maintain its role as the benchmark of international and universal human rights. It plays an important role in aspirational terms — the world's nations working together — and that is not to be lightly set aside.

There is a need for further internal reorganisation of the UN in the light of its developing policies and the increasing public interest in its work — largely through the consultation procedures and NGO presence at its more recent conferences. These have acted as powerful catalysts for the growing international co-operation of NGOs assisted by more sophisticated and accessible technology such as the Internet. While this co-operation has yet to rival the power of the transnational companies, it is nevertheless an important step to developing mutual understanding between different NGO sectors who are now being accorded a formal status in some parts of the UN's internal deliberations. This change also legitimises joint working at the national level to develop policy on international issues, increasing their political influence on the domestic agenda and the development of co-operative international campaigns.

One specific issue of growing importance is whether it is right for the UN to intervene militarily (through 'peace-keeping forces') in a country's internal affairs when there is a risk of genocide such as we've seen in ex-Yugoslavia and in Rwanda and Burundi. I believe it is, provided it is a decision made by the UN and with no vetoes allowed, because the human cost is otherwise too high and the wider, long-term repercussions appalling, not just within the state but for neighbouring countries faced with massive influxes of refugees, often leading to an export of the civil war.

We have seen that many countries — our own included — are not willing to take refugees and others find it difficult to cope, but the support offered to countries with huge refugee problems is not sufficient or consistent. We know mass displacement of people comes at enormous personal and environmental cost. The world's rich countries don't want migrants, yet won't adequately support the countries to which they flee and will not seriously work to reduce the conditions which force them out. That's why Greens are so passionate about the need for change.

The future for Europe

I've already looked at our defence programme and its development within the OSCE but, for most people these days, 'Europe' means the half with the European Union.

I want to make it clear that we're 'in Europe' and that 'what they do in Europe' includes us in the UK. I hate the language that separates us from our geography, a large part of our history and a considerable part of our cultural and political past, present and future. Whether we stay in the European Union is another issue. 'Europe' is more than 15 countries.

The EU is trying to do too much. I don't want to see a United States of Europe, in the sense of having an over-arching government, its own currency, bank, constitution, citizenship, armed forces, legislative system, etc. — all the indicators of statehood. But I can understand those who feel that a cohesive structure is needed, binding states together in common cause and pooling some of their sovereignty so that they will never go to war again. I don't agree that we need a complex, federal unit to do it: the UK Green Parties don't support federation but some other Green Parties in Europe do.

What I want to see is European co-operation and common agreement on issues concerning the environment and human rights which covers a number of economic issues. I personally like the idea of the free movement of people within the EU and the concept of a European citizenship, giving reciprocal rights to EU residents (not nationals). I see this as part of the development of a wider, more global, citizenship.

Greens want to see the enlargement of the EU to embrace those countries, or autonomous regions, which wish to join and express that through a referendum. Those areas must meet basic democratic standards. But I don't want to see them having to subscribe to the total free trade requirements of the Single European Act nor having to implement the current Common Agricultural Policy (CAP).

There's a clear problem in the EU resulting from a lack of public consent to political plans which are agreed by governments who have not consulted the people. Let us have none of this stupidity about 'rule from Brussels': decisions within the EU are made by politicians. It is governments, not bureaucrats, that are failing to consult.

Economic policy in the EU is being driven by conservative dogma. The convergence criteria for Economic and Monetary Union (EMU), dealing with inflation, government spending and borrowing levels, are being used either as an excuse for cutting back on social spending and people resent that.

However, people do want the EU to deliver on environmental and social protection.

The Green Group in the European Parliament opposed the Maastricht treaty because it failed to make any significant difference to these areas and perpetuated the 'Europe of Big Business'. The Green Party recently decided that Britain should withdraw from the EU because it is still moving in that direction and seeking greater integration.

At the same time we should also recognise that a lot of our 'problems' with the EU are to do with the British system of government. We must have elections to the European Parliament under proportional representation so that we are represented on the same basis as other countries. Such elections would also mean that the political balance of the parliament is not distorted by our disproportional system.

We need better reporting on the EU: to have MEPs appearing regularly on Question Time would be revolutionary. We must have referendums on issues connected with European sovereignty, such as the outcome of Inter-Governmental Conferences. We are as capable as the French and Irish of understanding the issues — maybe our politicians can't explain them.

There should be better links between Westminster and the European Parliament. This would encourage debate and help recognise the battles being waged within the European Parliament which are barely debated at Westminster. These include genetic engineering, energy taxation and policy, agricultural policy, police co-operation and relationships with Central and Eastern Europe.

I believe that the joint experience of different political cultures is extremely valuable. Personally, I believe that the issues which are shaping the future are taken far more seriously at the EU level than they are domestically and that increasingly people in the UK (if not Westminster and the Home Counties) do feel a sense of identification with Europe. The value given to issues of sustainability is, of course, due in part to the presence of directly elected Greens who have done much to ensure they are discussed. I am not yet ready to leave them to it. I want the UK to go through the same educative process and to do it openly.

I want an ecologically sustainable European Union where there are agreed frameworks and targets for, say, the reduction in pesticide use or the increase in electricity provided by renewable sources or a timetable to phase out nuclear energy. A Union which will develop a conversion programme for the arms industry; lay down standards for transnational companies operating within its boundaries that would have to be adhered to globally; work together to reduce carbon dioxide emissions; and provide research findings and assistance to the poor world on habitat protection. I believe such things are achievable at the EU level: I don't think a lasting monetary union is, particularly in an enlarged Union.

The battle for the direction the EU will take — grey or Green — is not yet over: it is a battle which has barely begun at Westminster.

In the longer term, Greens want to see a Europe of the Regions, recognising that in many parts of Europe as a whole people feel a stronger regional identity than a national one. Greens also believe that the regional level of government is closer to the people and a more appropriate level for greater economic development and management of the environment. It therefore makes sense to develop the role of the Council of the Regions within the EU structure as a step towards such a Europe, leading in the medium term to possible co-decision making with the EP, replacing the Council of Ministers in the current set-up.

The future architecture of the EU is uncertain. Many Greens want a Confederation but could a Confederation, in the traditional terms of one area/one vote and a right of veto, really work?

What will be the tensions if the members of a Confederation of the Regions are directly elected, rather than representing their governments? Will it have legitimacy with the public if it is a body of appointees?

Whatever the difficulties, the concept of a Europe of the Regions, closer to the people, giving equal weight to all areas and expressing a desire for progress by consensus, is well worth pursuing.

· 11 ·

Getting the ecology right

Greens have often said that the only parties who have 'environmental' policies are grey parties. In other words, if you see issues as interlinked, in a sustainable way, then you are an ecologist and a Green. If you can pull out just one strand and deal with it in isolation, then you are an environmentalist. As I've already shown, environmental protection has as much to do with economic and social values as with caring for trees and green space.

When environmental issues were beginning to have an impact on conventional politics there was a rush to develop 'environmental policies' and parade them to prove Green credentials. Greens were sceptical, and to some extent still are. 'How much are you prepared to change to protect the environment?' is still a telling question.

It is true that there are changes going on: at least there is now some debate about eco-taxes, 'good' food, air and water quality where before there was virtually none. There is a growing recognition that there are jobs to be created in environmentally useful work but the breadth and depth of action needed is not really comprehended by many.

We have to understand that our lives are limited by what this planet can provide, and to beware the siren voices that talk about using space travel to mine other planets for what earth lacks — the assumption that those raw materials exist, and that we will have the technology, resources and time to find them and exploit them in sufficient quantity is truly staggering. Anyway, for all we know, other life forms have already done it and are now extinct.

There are also advocates of deep sea exploitation of resources. To some extent that's already begun with offshore oil and gas extraction helping us to stave off the time when we really have to live within our planetary limits. But the cost of extraction is not cheap, and neither is the technology: indeed, we are still running up against the pressure problems involved in exploring the oceans. At some point in the distant future, there may be an argument for further exploitation. However, knowing what we now know about the environmental risks of excessive exploitations and our growing realisation of the importance of the oceans to our climate, we would be very unwise to rush to treat them as unthinkingly as we treat the land.

Indeed, the search for other rich sources of materials so that we can continue our profligate use of resources, only displaces the problem of reaching the limits, rather than dealing with it constructively.

Conserving resources

A constructive approach means we must conserve our resources, minimise the damage we cause, restore what we can and look for positive alternatives: we have to change our attitudes and expectations.

This means concentrating on meeting real *needs* globally, rather than addressing the consumer-driven 'wants' of a minority.

We cannot expect to sustain the material living standards presently enjoyed by the better off in the USA, Europe and other predominantly industrialised countries. We might hope to maintain the 'bike and bus' economy of many in Europe and elsewhere, and we must ensure that the living standards of those who are today the world's poor rise to meet their needs.

This future will depend, amongst other things, on our numbers. The more we are, the less we can expect as our share — however fair the distribution system. There could well be a time, some predictions say sooner rather than later, when our sheer numbers will make sustainable living impossible, so an overall limitation of our population is essential. That is why family planning programmes are necessary and why it is critically important that women should have the means to control their fertility. China has already decided that women do not automatically have the right to bear children, and that a person's right to have a family is subordinate to a collective right to have access to the necessities of life. If we do not wish to see more such enforced programmes in future, we have to find more positive ways to persuade people to have fewer children and accept that childlessness is not a failure but is often a morally responsible choice.

In order to conserve our environment, we have to manage its resources as efficiently as possible, resources such as fossil fuels, metals, minerals and land for agriculture and as habitat.

One of the most effective ways to ensure we use our finite natural resources efficiently is to tax them so that consumers realise their finite nature: the more expensive they are, the more likely we are not to waste them and to re-use or recycle them efficiently.

A lot of our metals, especially those in domestic use, end up in landfill sites or in incinerators. Yet the energy gains from recycling such products can be enormous. On a global basis, an estimated 30% of the aluminium we use is recycled — it could easily be 80%; the energy gain (that is, the difference between energy which is used in smelting new aluminium and in recycling) can be 95%. Even without a resource tax, its worth recycling to the extent that some schemes in the USA encourage homeless people to collect cans in exchange for their scrap metal value, providing at least some income. I've seen it done in my own area on an informal basis. With a deposit charged, and/or a resource tax on virgin aluminium ore, recycling would be even more worthwhile.

The savings for other metals and products such as paper are not quite as dramatic (about 50% energy savings on iron at present) but they are considerable. With a resource tax levied at the point of extraction, or import (if there's no equivalent domestic tax), there would be a considerable incentive for efficient use and recycling or, where appropriate, re-use.

The UK has one of the most efficient re-use systems of glass bottles anywhere in the world through our doorstep milk deliveries, now in decline due to supermarkets employing one-use containers (there aren't even many recycling schemes for the plastic ones) and cutting prices to attract buyers. A resource tax would make the bottle delivery system more attractive as it could carry a lower rate for re-used products. That would also help to keep what is also a social service in many respects.

Pricing differentials would also help stimulate the market for recycled products — a log-jam at times — and help us reach more ambitious levels of recycling domestic waste — say, 60% of recyclable waste in 5 years.

Another method used in some states in the USA is to set minimum requirements for the amount of recycled material used in certain products — paper, for example. This has also been combined with tax incentives or grant and loan schemes to ensure effective local markets.

A tax on packaging too would help reduce the resources used for this, putting less into the recycling bin. Germany and Switzerland already have schemes where people pay for the amount of waste they have collected; consumer pressure has therefore played a considerable role in prompting producers to reduce their packaging.

Taxing energy

For some time, the EU has been aiming to introduce an energy tax on fossil fuels (the UK has voted against it every time). There are a number of good reasons to reduce consumption of fossil fuels: to make them last over a longer period; to reduce carbon dioxide emissions and to reduce other pollutants such as acid rain and low-level ozone. In the fuels burned for electricity production, there's a considerable loss of energy of up to 60%

between the fuel entering a power station and what we get once we have turned on the switch, so there is enormous room for improved technologies to make a difference. We are currently insisting on using natural gas as part of our overall electricity generation strategy when its most efficient use is when burnt for heating in homes or industry. Raising transport costs also helps to make local goods more competitive and can help local markets.

When putting forward their proposals for the energy tax, the Green Group in the EP wanted the tax to serve a definite purpose: they wanted it employed so that it did not simply encourage switching to a lower-rated fuel or to nuclear power (an unacceptable long-term risk for Greens, energy intensive in the building of plant and inefficient in delivery) but to conservation, efficiency and renewable sources. So the Green Group's tax had a dual purpose — to tax carbon dioxide emissions, in order to help reduce global warming, and to tax the fuel's overall energy value (this latter element of the tax could be varied to cover other environmental factors). The Group proposed a $2/3$ carbon, $1/3$ energy split.

The tax would be incremental, rising at a given rate over 8 years. On their projected figures in 1992, they felt the measure, and consequent actions to decrease fuel use, would lead to a 20% reduction in carbon dioxide in the EU by the year 2000, rising to 50% by 2010 before eventually reaching the required global target of at least 60%, but nearer 80% if adjusted for the North-South emissions imbalance. Energy taxation is only one of the measures needed. (Our government's aim at the time was for the stabilising of carbon dioxide emissions at 1990 levels by 2005).

Revenue from the tax would do five things:
• help compensate low-income households for the resulting price rise;
• reduce the taxes on employing people;
• fund an energy conservation programme;
• provide additional research and development on renewable energy sources;
• provide assistance to East Europe and poor countries to move to ecological energy systems.

Of course, fuel can also be taxed according to its use, so it is possible to tax petrol at a rate that makes using a car even more expensive and thus discourages unnecessary use or even ownership.

Preserving diversity

However, there are other natural resources we should be conserving — variety of habitat and overall bio-diversity. Most species are destroyed by human action, and one of our activities is destroying habitat by building on it, cutting it down, or moving in and taking over. If we are to maintain the vast range of species and the genetic diversity within those species, we have to learn to live with that habitat. Reducing our numbers will take off some of the pressure; paying a fair price for goods means you can, say, cut fewer trees or plant fewer acres for the same return, and reforms of land ownership can help keep people in place. But all this is not enough.

In some parts of the world, we have to look at how indigenous peoples have lived and how others have worked with their environment without destroying it — rubber extraction and harvesting forest foods such as nuts in the Amazon, coppicing forests in parts of Europe, or developing the use of traditional medicines as part of public health care as in Thailand and Nepal.

Eco-tourism is being promoted as a way of encouraging local people to make a living from conserving wildlife and capitalising on the growing interest in seeing it where it belongs, and not in the artificial collection scenario of zoos.

There is some money available, but not yet enough, through the Global Environment Facility (GEF), set up by the UN in the light of the Rio Summit. This derives money from the World Bank, the UN Development Programme and the UN Environment Programme and is for investment in preserving the 'global commons' — the parts of the world considered to belong to us all (which are the atmosphere, waterways and bio-diversity) and a number of projects are already being funded.

However, at least we need to be devising a way of having a range of strategically important sites protected as genetic banks or as water catchment areas and this means working closely with local people if it is really to work.

The vast tropical and temperate rainforests are the 'lungs of the earth' and crucial to the maintenance of weather systems: forests are often at riverheads and they play an important part in retaining soil and are potential 'sinks' for excess carbon dioxide. Their maintenance is essential in helping to slow global warming.

How can they be sustained? I've already mentioned harvesting products but there is growing interest in agro-forestry or growing crops alongside trees in the right combinations to improve soil quality and meet local needs. Using already logged areas as sites for maintained woodland, rather than cutting further into primary forests, is also being developed — in the USA as well as elsewhere it can increase production and save the remaining original forest.

Growing more of our own wood for our own needs would also help. It would significantly reduce imports, provide work and help maintain and develop a diversity of habitat at home. We're gradually moving away from stands of firs to mixed forests which can encourage a variety of uses and a greater diversity of flora and fauna.

The government is supporting the development of community forests — joint enterprises for work and recreation. But we would go further: we believe that we should have a general 'right to roam' the countryside, except in very restricted circumstances, such as protecting wildlife during breeding seasons.

We can also plan SSSI protection and more varied habitat into our own developments in both towns and countryside. Indeed, by changing our planning system so that public inquiries look at all possible options when a significant proposal is submitted and ensuring funding for objectors, we could improve the quality of the whole planning debate in this country. There should be criteria for what constitute proper environmental and social assessments so that these really aid the process and are available to the public under Freedom of Information legislation.

The genetic threat

We need to resist the temptation to turn increasingly to genetic engineering as a substitute for species loss. It has been said that genetic engineers could have an impact greater than that of atom scientists — and that worries me. The bio-technology market is potentially huge and it could also lock people into a relationship with transnational corporations which does not increase their choice or necessarily provide sustainable agriculture. At the moment, we don't fully understand the consequences of all our experimentation for our ecology.

It was put to me at a seminar attended by senior people from a number of big agro-chemical companies that they are at a junction and can either go down the bio-technology route or follow the integrated environmental management road. They were waiting for a clear signal.

What are they getting? Rio and the Bio-Diversity Treaty, which doesn't treat bio-diversity as a 'global common' but as national property (despite the GEF) which still ought to be protected, and GATT with TRIPs (Trade-Related Intellectual Property Rights) and talk of a protocol on bio-technology in the future. Meanwhile, in the EU, the fight continues — led by the Greens — to prevent the patenting of genetically engineered material such as the Neem Tree from India (24 patents in the USA at present) and a cell line from the Guaymi people of Panama. Greens are totally opposed to patenting life, from whatever species, and want the ethics of genetic engineering fully debated in public. The precautionary principle

— don't carry on if you don't know what the outcome will be — should apply to this area as much as any other.

The reform of agriculture

The principles of conservation and damage limitation meet in the way we farm. Large-scale monoculture has taken its toll of the land and we need to change. Greens have always advocated an integrated approach to farming so that animal waste products can be used as on-site fertiliser rather than going into slurry ponds and vegetable waste can be used as animal feed: the greater use of organic matter as fertiliser increases soil quality and helps reduce erosion. We want to see less intensive production, requiring fewer artificial inputs and eventually leading to organic agriculture. We are encouraged by the developing interest in integrated crop management which aims to make the best use of crop rotation; selects seed varieties with the best natural resistance to pests and diseases; uses the lowest possible inputs of artificial chemicals and fossil fuels; aims to increase soil quality and takes account of wildlife and landscape considerations — so that hedges and other 'wildlife corridors' are re-introduced or maintained. We also want to see the development of clear and widely understood labelling programmes so that people know how their food is produced: it's interesting that people put their faith in a label supported by independent, voluntary organisations who have filled an enormous vacuum left by government. I want to see the day when food produced by chemical agriculture takes up a small amount of shelf space and has to be labelled and we assume all other produce to be organically produced!

Factory farming has to go — it's cruel, unnatural and encourages a whole chain of problems suppressed by drugs, or by even more bizarre feeding and housing regimes. We should at least be eating much less meat — some would argue we should be eating none at all — and only keeping animals that we can feed without importing grain or resorting to processed animal waste.

The best way to improve farming methods and our confidence in food production is to develop closer links between producer and consumer and encourage local markets. This could be helped by dismantling the CAP and its price support and intervention buying, and using some of the money to support farmers' incomes (through, say, a Farm Revenue Stabilisation Scheme) and to keep farmers on the land by diverting other funds to supporting rural development (through Regional and Structural Funds if you want to keep the European Union involved). This could then ensure farming is seen as part of a way of life — putting the 'culture' back into agriculture — not separate from other work in the countryside, whether forestry based, small-scale industry or home-working (of the better-paid sort). Local co-operatives and other marketing strategies could be supported. Without the CAP, and by introducing an energy tax so the price of imported goods would rise, we could also bring more land back into production: we could also do away with the 70% of organic food we import at present and grow it ourselves.

A shift in the producer/consumer relationship, lessening the power of the large retailers, would also help in the development of smaller-scale farms. Greens want to see farms run by farmers, for the public — not by managers (however good they may be individually) for absentee owners (like pension funds) and primarily for profit. We want to see the large landholdings broken up. The Ecology Building Society was set up originally by Green Party members who wanted to find a way to help those who wanted to buy smallholdings (not, at that time, willingly lent on) and to help increase their number.

Small-scale farmers feel themselves constantly under threat and the rate of those leaving the land is high, yet these are people who want to live in the countryside and generally protect it. They are potentially important allies in the quest for sustainability and should be supported.

A pattern of farm ownership that is becoming more common in places such as Italy is a mixed-work format: people live on the farm but earn their income elsewhere and thus combine with those who work basically as the primary producers. They might share in cheese or wine production and the marketing; holidays will be taken to coincide with harvest and the paperwork might be done by outside workers. This provides support for the farmer — they get company, the possibility of leaving the farm for breaks, removing the isolation they can otherwise experience. It also keeps people on the land. Italy is also consciously developing its agro-tourism industry, not just as a way of providing income, but also to help deepen public understanding.

So more organic, less intensive, less animal geared agriculture would do a better job of conserving healthy, fertile soil, produce less slurry and chemicals to run off into our water supply and use less energy-intensive inputs. We could further encourage the trend by a nitrogen tax.

By reducing field size, introducing tree breaks and hedges we can also improve bio-diversity and reduce soil erosion.

Soil quality could also be improved by using new irrigation systems, which take water directly to the parts of the plant where it's most needed and can be three times as efficient as the old systems, and produce less salination and fewer drainage problems. To replace old systems will take time and capital but would be a sound use of World Bank cash, increased aid programmes and prove a wise government investment in countries like the USA.

Cleaning up the water

Changing agricultural and forestry practice is just one way of reducing the damage we do to our water supplies.

Another is to reduce the pollutants we put into our rivers and seas: this would result in our spending less on cleaning up water — a difficult process. Some progress has been made, but it is still insufficient. Sewage is not adequately broken down before release into the sea, so we still expect the oceans to act as cleansing plant. There is a need for greater investment in new plant and new systems. The aim must also be to have as little pollution going into our waterways as possible. While a greater risk of prosecution and higher fines is one worthwhile way of trying to deter problems, it is not sufficient, even if the regulators are adequately funded and staffed.

There is a need for more information about what goes into manufacturing processes and what comes out and certainly more research into the effects of combinations — which are not always tested for. Indeed, the regular testing of sea water does not test for the vast range of chemicals put into it. We have to try to keep such pollution out of the water cycle.

There are over 100,000 substances on the inventory of chemicals on the EU market (there has been a massive increase in the last 20 years) and toxic waste is recognised as a by-product of economic growth. Even the EU does not have sufficient facilities to deal with all it produces — hence the iniquitous exports to less regulated countries. Having waste classed as a 'good' (i.e. a product) within the EU has meant it has not been easy to keep track of its travels. There is a real need for international co-ordination on chemical testing and research findings. The International Programme on Chemical Safety was identified at the Rio Summit as the nucleus for such co-operation and should set standards and provide public information.

The overall goal, however, must be to reduce the use of chemicals and improve clean-up techniques for those we must use. This will need a change in philosophy requiring manufacturers to be responsible for the whole lifecycle of their products and by-products, in whatever country they operate. Having to pay for all their activities provides a powerful incentive to reduce harmful processes. The penalties for illegal dumping must be severe. A change in insurance liabilities could concentrate the mind! Countries should have to deal safely with their own toxic waste (radioactive or otherwise) rather than export it: help should be provided to countries such as Romania and Nigeria which have acted as cheap dumping grounds in the past.

Expensive remedies

So, restoring damage done is a tough task — better not to cause the problems in the first place. We need to be diverting monies from military research and development into looking at better ways to clean up our environment, so that we can trust our water supplies and use land for housing, agriculture and recreation with confidence. It will require restraint and expenditure.

If we are to prevent the exhaustion of our fish stocks, for example, we have to consider total bans on fishing in certain places (the fallow box concept) or in certain seasons. I would start with reducing the number of factory ships. Many of them are unsafe to be at sea and could be impounded on these grounds alone. With a 'Green Agreement on Sustainable Trade', we might have grounds for boycotting certain Japanese goods unless Japan reduces whaling and factory fishing. An international system of fishing zones, with certain types of fishing banned at certain distances from the coast, would help protect coastal fisheries and local communities.

We also have to develop our use of renewable resources to replace other resource use wherever we can. Renewable resources also have to be used carefully: the same concerns about waste generation, pollution and conservation before consumption apply to renewables. Even if we have recyclable cars powered by hydrogen engines, they don't solve the problems of congestion and overall consumption levels. Solving one part of a problem is useful but you only get the full benefit when you've solved the entire problem.

Apart from increasing the use of renewable materials such as timber or straw in our buildings or growing sugar or corn to provide a replacement fuel for petrol in vehicles, the main area for the exploitation of renewable resources is in energy, providing power for heating, lighting, transport and industrial processes.

The UK attitude to such things has generally been half-hearted, to be polite, particularly when compared to countries such as Denmark. In 1992, such energy supplied about 2% of our electricity. Government projections for 2025 show renewable energy as supplying only between 5% and 20% of our electricity supply. We have only one wave-powered generator in the UK (at Islay), despite being an island in tidal seas.

Plans were announced by PowerGen in August 1996 for an offshore windfarm near Great Yarmouth, projected to produce 1.5 megawatts of electricity. There are small-scale solar energy deployments in the UK at a more domestic level although passive use of sunlight contributes about 15% of the UK's space heating needs.

Estimates for future capability vary: as technology develops it may be possible for the UK to supply 20% of its electricity needs through wind power and more again through wave power. Solar also has a part to play through a variety of technologies from parabolic dishes to photovoltaic cells.

Critics of renewable sources will often argue that the amount of land (or sea) needed to produce the equivalent power to that from conventional power stations is too great; that the technologies are untried and that the systems are ugly. The aesthetics are a matter of debate: personally, I wouldn't give a beauty prize to Thurrock power station or Sizewell A or B.

The technologies are still developing but they have been underfunded for years as nuclear power has been given preference for state funding. Many experimental projects for renewable energies have been funded by enthusiasts and voluntary groups: even some of the implementation programmes are funded in this way.

As to the geographical impact, the answer is — it depends. Certainly, if you are wedded to the idea of large-scale production, there may be some truth in the argument, but that is not the only way it can be used. The landtake for conventional power stations can be considerable too.

The beauty of many renewable energy sources is that they can be used on a variety of scales in a variety of situations. Passive solar panels can help heat domestic water supplies. Photovoltaic cells can provide on-site power directly, cutting out the need for grid connections: they are beginning to have a considerable impact in poorer countries such as Zimbabwe and the Dominican Republic as well as being used in pocket calculators and remote lighthouses.

Windpower can be used to provide power on single farms or to be added to the national grid system (an Ecology Party candidate in 1979 advocated using windmills on tower blocks!). Wave power can also be used to provide power on a small scale to island communities as on Islay or on a larger scale to supply the grid.

There are other methods of generating electricity, too, such as hydro-electric, tidal barrages or geothermal power, which can all make a contribution to cutting carbon dioxide and other emissions and helping to conserve fossil fuels for other uses.

What will be needed in future is a system which builds on reducing overall need for energy, greater conservation, increased efficiency and then a range of power sources to meet different needs. That can't be done without an integrated energy strategy, working alongside other changes. That strategy needs industrial development and public consent if it is going to work: those sectors need to work together and it is the role of government to achieve that. In the UK, the residual powers of the old Department of Energy nestle under the wing of the Department of Trade and Industry — quite the wrong way round, as energy use underpins the future of trade and industry, not vice versa.

If real change is going to happen, we need to change government priorities, whether through having ministers for energy and natural resources, strengthening the powers of local and regional government, increasing access to information, really involving people and so on...

How can we do that?

· 12 ·

No change, no chance

'A changing world demands a change of politics'. This was the message on an Ecology Party poster in the 1980s and it's still true.

Change is on the way. We are beginning to move towards a more ecologically sustainable society. There are many encouraging signs: people out to 'Reclaim the Streets'; students wishing to invest their future in developing ways to protect our environment; people at the grassroots working within LETS and other schemes to provide working alternatives, and so on.

But at the national political level, there is still so much to be done. However much a tiny minority fight for even some improvement, there are the many dead hands who want to minimise change or who don't see the problem. The Treasury controls the government and the free trade, de-regulation, low tax credo dominates almost all decision making. I find it hard to believe a change of government will make any real difference when I hear a Labour front-bench speaker say things like, 'The growth dividend will pay for an integrated transport system.' This is really a new way of saying that 'something will turn up'! A promise to use speculative money to fund a core policy is not convincing, particularly when the premise on which it is based — that conventional economic growth is the answer — is actually part of the underlying problem, itself undermining any attempt to create a sustainable future.

So how do we get change? How do we exchange the grey politicians for Green ones?

The usual method is the transformation route, whereby you seek to change the opinions of those already elected or standing for election. It's the route of public meetings, lobbying, letter writing, putting people on public platforms with those whose opinions you want them to hear, etc. It says to politicians — people care about this and may vote according to your response. It can work and is a very useful educative process for all involved. The Home Energy Conservation Act came into being this way.

But such a campaign can also be dismissed as the efforts of only a comparatively few though well organised people. Or the politicians being targeted may simply disagree (an entirely honourable reason for refusing to go along with what is proposed). Or it may clash with key parts of party policy or ideology — although this doesn't stop some! The transformation route is also a very slow one.

It would be far more satisfactory to vote for the party with the core values and policy positions that support a truly viable future. When Greens are elected, things change and the pace can accelerate. As Claire Joanny, elected for Les Verts in the EU elections in France in 1989, put it: 'When the Greens walked into the committee room, the others sat up'. Why? Because they knew they would have to defend their political positions and decisions — not just to a couple of Green MEPs, but to the millions of people who had put them there.

The election of those members has a wider effect too. It allows those in other parties who may be sympathetic to argue the case for change within their own movements. They can justify change, not just from the logical and moral position, but from a pragmatic position — there are votes in it.

The pressure groups find they have more than one route into the decision-making process. They can use this to maximise a progressive form of competition within the institution but they also have more ways in which to bring information out. It also means that they no longer look like 'clients' of a single party, which can help them to maintain a more independent stance.

The public, too, begin to see how things might change. If those who are elected do a competent job, this should encourage more people to vote Green next time.

In the UK, of course, with our disproportional electoral system, there are more barriers to overcome. All Greens, and many in some other parties too, are familiar with the 'wasted vote' syndrome — all the excuses that go, 'Well, I would vote for you but you won't get elected/won't change the government/it's important to get this lot out/one of you won't make any difference.' It's even more frustrating when prefaced by 'I agree with everything you say, it's such common sense, but ...'

I can think of few better reasons to change this system than the fact that it makes people feel they have to vote for something that they don't believe in and which doesn't make sense because they think it's all they can get. It's one of the reasons I spend so much time working for the introduction of proportional representation. People should be able to vote for what they believe in and, if there are enough of them, see their views represented.

But I know that true change needs more than a vote.

Our lives are made up of millions of decisions which shape the future of our planet. Our own decisions about the way we live really are important.

But the framework in which we make those decisions is decided, ultimately, by governments. They can control transnational companies if they choose to; they can change the basis for international trade; they can decide how much power local communities have; they can decide to bomb the world to pieces.

Increasingly, governments have power because we, the people, wish it: by voting we choose who will make those decisions. Yet in many countries there are people dying for the right to vote and to have more control over their futures.

What sort of future?

Without fundamental change in the way we live and in our attitude to the planet that sustains us, I see no chance of a just and peaceful future, either within this country or on a global basis.

Change starts with ourselves, the values we hold and the actions we take. Governments reflect the values of those who elect them. Our actions at the ballot box are as important as any.

In the *Manifesto for a Sustainable Society*, first written in 1975, it says:

> The search for a sustainable society is a new venture: the Green Party is quite clear on the new direction we must take, but no one can possibly know all the details at the outset ... The Green Party offers not a panacea, but a foundation for a new way of life.

Together, we need to continue building the foundations if we want the chance of a safe, secure future.

Bibliography

Manifesto for a Sustainable Society   Green Party; updated regularly in the light of Conference decisions

The New Protectionism   Colin Hines & Tim Lang, Earthscan, 1993

LETS Work: Rebuilding the local economy   Peter Lang, Grover Books, 1994

Ethical Investment: A saver's guide   Peter Lang, Jon Carpenter Publishing, 1996

Measuring Sustainable Economic Welfare – a Pilot Index, 1950-1990   Dr Tim Jackson & Nic Marks, NEF and Stockholm Institute, 1994

Democracy and Green Political Thought – Sustainability, Rights & Citizenship   Brian Doherty and Marius de Geus (editors), Routledge, 1996

Greens in the European Parliament – a new sense of purpose for Europe   Diana Johnstone, Green Group in the EP, 1994

The Politics of the Real World   Michael Jacobs for the Real World Coalition, Earthscan, 1996

State of the World Report   edited by Lester R Brown, Earthscan, published annually

Green Political Thought – An Introduction   Andrew Dobson, Unwin Hyman, 1990

Index

Accountability, 6, 26, 86
Acid rain, see Pollution,
Added value, 15, 38, 51
Adult education, see Education
Advertising, 76
Africa, 28, 40, 45, 47, 48, 69, 78
Afro-Caribbean, 9
Age Concern, 79
Agenda 21, see UN
Agriculture, 48, 49, 52, 67, 68, 100, 104, 105-107, 108; agro-chemical, 104; agro-tourism 106; CAP, 51, 96, 106; cash-crops, 36, 39; factory farming, 52, 105, ; monoculture, 52; organic farming, 67, 86, 105, 106; organic food, 106; McSharry reforms, 51
Agro-forestry, see Forests
Aid, 38, 88, 90, 93, 104; bilateral, 93; multilateral, 93
AIDS, 69
Air quality, 84
Alcohol, 79, 84
Aluminium, 100-101
Amazon, 15, 103
Animal rights, 41, 75,
Antarctic, 43, 46
Anti-racist, 77
Aquifers, 48
Arab, 34
Aral Sea, 49
Arms, 31, 32, 34, 35, 89, 90
Asia, 9, 28, 40, 45, 47, 48, 63
Association for the Conservation of Energy (ACE), 66
Asthma, 19
Asylum, 33, 77, 78
Audit, Democratic, 27

Bangladesh, 36, 38
Banking, community, 63
Basic Income Scheme, 60, 61, 70
BBC, 13
Bill of Rights, see Rights.
Bio-diversity, 18, 40, 42, 52, 102, 103, 104, 106,
Bio-regions, 75
Bio-technology, 104
Brazil, 36, 38, 51
Brundtland, 41, 54

BSE, 51
BST, 50
Buses, see Transport
Bush, President, 32, 42
Business, 11, 15, 22, 32, 42, 51, 63, 65, 70, Asian, 63, black, 63, community, 70, failures, 11, small, 11

California, 20
Cannabis, see Drugs,
CAP, see Agriculture
Capitalism, 31
Capitalist, 32
Carbon dioxide, 45, 98, 101, 102, 103, 109
Care, 8, 20, 25, 78, 83, 84, 86, child care, 61, 62, health, see Health
Care and Repair, see Housing
Cars, see Transport
Cash crops, see Agriculture
Central Europe, 90, 92, 97
CFCs, 43, 44, 45, Montreal Protocol, 44
Charter 88, 85
Chemical weapons, 89
Child care, see Care
Child labour, 37
Child Support Agency, 9
Children, 8, 9, 19, 20, 21, 22, 23, 37, 39, 40, 52, 59, 61, 71, 74, 80, 81, 86, 87, 93, 100
China, 100
Citizens Advice Bureau, 11
Citizens Army, 91
Citizens Charter, 26
Citizens Defence, 91
Citizens Initiative, 77, 86
Citizens juries, 86
Citizenship, 22, 77, 96; Foundation, 82
Clarke, Kenneth, 54
Climate change, 44; Framework Convention on, 45; Global warming, 42-46, 52, 56, 102, 103; Greenhouse effect, 44, Greenhouse gases, 46
Combined heat & power (CHP), see Energy
COMECON, 92
Common Agricultural Policy (CAP), see Agriculture
Communism, 32
Communist, 28, 38
Community, 3, 6, 19, 23, 24, 29, 30, 58, 60, 62, 68, 73, 74, 79, 81, 82, 84, 86, 93, 108, 109, 133; banking, see Banking; business, see Business; care, 25, 59, 83, local, 15, 25
Community Charge (Poll Tax), 2
Community forests, see Forests
Community Health Council, 78

Community service, 82
Companies, agro-chemical, 104; gas/water, see Utilities; nuclear, 32; transnational(TNCs), 16, 68, 69, 97, 104, 133
Complementary medicine, 84
Compulsory Competitive Tendering (CCT), see Housing,
Conservation, 6, 30, 41, 54, 57, 94, 100, 102, 103, 105, 108, 109,
Conservative(s), 6, 16, 42, 64
Conserver economy, see Economy
Constitution, 26, 29, 32, 75, 76, 77, 96; written, 28, 77, 78
Consumer economy, see Economy
Consumption, 6, 7, 39, 46, 47, 50, 58, 64, 92, 93, 101, 108, conspicuous, 8
Conventional Forces in Europe Treaty (CFE), 89
Co-operatives, 70
Council houses, see Housing
Council of Europe, see European Union
Council of Ministers, see European Union
Council of the Regions, see European Union
Credit, 9, 11, Family, 59; guarantees, 89; unions, 62, 63, 81
Crime, 3, 20, 37, 79, 81, 82, 83; offenders, 82; organised, 3, 33; street, 81; white-collar, 20
Criminal Justice Act, 29
Currency, single, see EU
Cycling, see Transport

Dafis, Cynog, 66
Death penalty, 37
Debt burden, 38, 39, 67; redemption, 63; relief, 93
Decentralisation, 6, 75
Defence Export Services Association, 89
Defensive expenditure, 58
Deforestation, 36
Delors, Jacques, 12
Department of Energy, 110
Department of Social Security (DSS), 9, 62
Department of Trade & Industry, 110
Dependency culture, 10
Desertification, 46, 51
Developing countries, 8, 37, 38, 44, 46, 90
Devolution, 6, 29, 84
Diet, 47, 50, 51
Disability, see Physical Impairment
Disarmament, 73, 90, 95; KONVER programme, 90; nuclear, 91; RECARM programme, 90; SALT talks, 31; START treaties, 89,
Drugs, cannabis, 37; trafficking, 33

Earth Summit, see UNCED
East Europe, 90, 92, 97
Eastern Bloc, 34, 87, 88
Eco-taxes, see Taxation
Eco-tourism, see Tourism
Ecology, 5, 40, 46, 54, 104
Ecology Building Society, 85, 106
Ecology Party, 53, 109, 111
Economy, balanced, 58; 'bicycle & bus', 100; casual, 13; compensation, 26; conserver, 57; consumer, 10, 57; global, 17, 39; green, 59, 64, 65, 67, 68, 70; informal, 33, 70; international, 15, 33, 69; local, 15, 62, 63, 69, 70; money, 7, 59; no growth, 11; regional, 15; sustainable, 59, 64, 65, 67, 70; 'tiger', 40
Education, 4, 14, 20, 21-23, 27, 32, 36, 37, 40, 60, 68, 71-74, 81, 84, 93; adult, 71; GCSE, 21; 'Gross Educational Product', 22; higher, 43, 74; National Curriculum, 40, 72; Open University, 71; Peace Studies, 73; school governors, 21, 27; sex education, 74, 84; students, 2, 9, 60, 71, 72, 73, 111; TVEI, 21
Elections, 30; European, 76, 97, 112; local council, 7, 29, 53
Electoral systems, 29, 77, 112; proportional representation, 75, 97, 112
Employment, 2, 3, 8, 12, 13, 64, 71, 79, 84; alternative, 67; casualisation of, 12; insecure, 22; of MPs, 41; rights, see Rights
Energy, 15, 16, 42, 46, 51, 57, 66-67, 75, 94, 100, 101-102, 106, 108-110; combined heat & power, 67; fossil fuels, 45, 46, 65, 66, 100, 101, 105, 109; natural gas, 32, 102; nuclear, see Nuclear; oil, 32, 38, 39, 69, 99; oil crisis, 1; oil reserves, 34, 46; solar, see Renewable Energy
Energy efficiency, 46, 69, 85; Insulation, 66, 85
Enterprise(s), 15, 63, 70, 104
Environment, Department of the, 4; UN Conference on the Environment & Development, 35; working, 42
Environmental, assessment, 104; audit, 94; costs, 51, 70, 95; damage, 31, 38, 58, 59, 68; destruction, 22; dumping, 69; effects, 36; groups, 31; impact, 66; impact assessments, 49; movement, 18, 31, 54; policies, 99; protection, 18, 42, 70, 88, 94, 97, 99, 111; rights see Rights; safeguards, 18; tax measures, 60,
Equality, 4, 10
Ethical investment, 17, 70; purchasing, 28; supermarkets, 70
Ethnic minorities, 30, 75; Afro-Caribbean, 9;

Index   117

Asian, 9; Asian business, *see* Business; Black business, *see* Business; Black men, 3

Europe, Conventional Forces in Europe (CFE) Treaty, 89; Fortress, 33; of the Regions, 98; United States of, 96

European Bank for Reconstruction & Development (EBRD), 33; citizenship, 96; elections, *see* Elections; Green Parties, 91, 96

European Parliament, 50, 68, 97; Green Group in the, 31

European Union (EU), 6, 12, 14, 35, 39, 47, 54, 91, 96, 105; Common Agricultural Policy, *see* Agriculture, Council of Europe, 92; Council of Ministers, 98; Council of the Regions, 98; European Commission, 12; European Community, 12, 22; European Monetary Union (EMU), 96, 98; Maastricht Treaty, 18; Single European Act, 96; Single European Currency, 16, 17; Single market, 68

Examinations, 72

Exports, 14, 36, 37, 38, 93, 107; arms, 90, 91

Factory farming, *see* Agriculture
Fair Trade, 68
Family Credit, 59
Family planning, 68, 93, 100
Fast food, 51
Feel-good factor, 12
Fishing, 42, 47, 108
Food chain, 44; co-ops, 63, 81; safety, 76
Forests/forestry, 36, 40, 45, 46, 52, 103, 106, 107; agro-forestry, 103; community, 104
Fortress Europe, *see* European Union
Fossil fuels, *see* Energy
Foyers, *see* Housing
France, 3, 15, 90, 91; French government, 36, 97, Front National, 3
Free market, 39
Free trade, 15, 18, 94
Freedom of Information, 76, 104
Fundamentalism, 5
Fundholding, *see* NHS

Gandhi, 56, 91
GATT, 6, 14, 17, 18, 39, 68, 94, 104; intellectual property rights, 18; trade-related intellectual property rights(TRIPs), 68, 104
GCSE, *see* Education
Gene(s), 52, 68, 103, pool, 40, 52
Genetic engineering, 50, 97, 104
Genocide, 95
Geothermal power, *see* Renewable Energy
Germany, 17, 34, 37, 66, 89, 101

Global balance of power, 40; citizen, 88, 96; citizenship, *see* Citizenship; commons, 103, 104; ecology, 40; economy, *see* Economy; relationships, 3; warming, *see* Climate Change

Global Environment Facility (GEF), *see* UN

Government; inter-, 93, 97; local, 29, 30, 41, 63, 75, 79, 85, 86, 110; regional, 75, 98, 110; unelected, 68; Westminster, 6, 84, 97, 98; Whitehall, 6

Green Group, *see* European Parliament
Green Party, 1, 7, 66, 106, 113
Greenhouse effect, *see* Climate Change
Greenpeace, 47
Gross Domestic Product(GDP), *see* Indicators,
Gross National Product (GNP)), *see* Indicators
Growth, 12, 18, 37, 39, 46, 54, 64, 70; non-inflationary, 70; sustainable, 11, 18, 54; *see also* Economy
Gulf war, 34, 38; Iraq and, 34, 35, 48, 87
Gummer, John, 111

Habitat, 36, 52, 98, 100, 102, 103, 104
Health and Safety, 65, 69
Health care, 25, 32, 40, 68, 83, 84, 93, 103
Health centres, *see* NHS
Hereditary principle, 75
Himalayas, *see* Soil Erosion
Holistic, 30, 54
Home Energy Conservation Act 1995, 66, 85, 111
Homeless people, 2, 76, 101
Homeowners, 9
Homeworkers, 12
Hospitals, *see* NHS
Households, single-parent, 9, 10, 61
Housing, 8, 23, 24, 25, 29, 32, 41, 58, 63, 64, 76, 84, 85, 105, 108; Associations, 9, 78, 85; Care & Repair, 66; Compulsory Competitive Tendering (CCT), 63; Council houses, 9; Foyers, 84; Housing Benefit, 9, 60; repairs, 69; sheltered, 83; social, 9, 16, 85
Human Development Index, *see* Indicators
Human rights, *see* Rights
Hussein, Saddam, 34
Hydro-electric, *see* Renewable Energy,
Hydrogen, 108

Immigration, 33, 78
Import substitution, 94
India, 63, 104
Indicators, 55, 83, 92; different, 58, 59; Gross Domestic Product(GDP), 12, 46; Gross National Product(GNP), 58, 59, 93, Human Development Index, 59; Infant

mortality, 8, 39; Sustainable Economic Welfare (ISEW), Index of, 58
Industrial revolution, 12, 14, 40
Infant mortality, see Indicators
Insecurity, see Employment
Insulation, see Energy Efficiency
Insurance, 12, 17, 43, 52, 107
Intellectual property rights, see GATT
International Monetary Fund (IMF), 36, 38, 39, 68, 93, 94
Internet, 95
Investment, inward, 14, 15; ethical, see Ethical
Irrigation, 48, 49, 106
Israel, 28, 48

Jail sentences, 20
Japan, 16, 38, 47, 48, 108
Job insecurity, see Employment
Judiciary, 76

KONVER programme, see Disarmament

Labour Party, 57, 64
Land, 15, 23, 31, 32, 36, 37, 39, 41, 44, 45, 49-50, 51, 75, 84, 86, 93, 100, 103, 105, 106, 108, 109; Land seizure, 5; Land Value Taxation (LVT), 86; Reclaimed, 24, 89; 'Right to Roam', 104
Landfill, 66; 100, tax, 66
Landmines, 88
Legal system, 81
Les Verts, 112
Liberal, 5
Liberal Democrats, 6, 61, 66
Literacy, 40, 59, 82, 93
Local Agenda 21, 85, 88
Local Exchange Trading System (LETS), 61, 62, 81, 111
Local government, see Government
London, 14; government, 29
Lorries, see Transport
Low waged, 13, 15, 61

Maastricht Treaty, see European Union
Major, John, 2, 26, 42
Malnutrition, 8, 77
Manifesto for a Sustainable Society, 1, 53, 113
Migration, 33, 38, 93
Military, 1, 31-34, 90, 91; expenditure, 5, 43; research and development, 108
Minimum wage, 61
Money markets, 16, 17
Monoculture, see Agriculture
Monopolies and Mergers, 69

Montreal Protocol, see CFCs,
Mortgages, 9
Mutually Assured Destruction(MAD), 91

National Curriculum, see Education
National Grid, 109
National Insurance, 60, 64
National Lottery, 9
National Trust, 41
Nationality, 77
NATO, 6, 35, 91, 92
Natural gas, see Energy,
Natural resources, 15, 36, 38, 40-42, 46, 65, 75, 87, 102, 110; finite, 100
Nature, 7, 22, 42, 50, 51, 54, 74,
Neighbourhood Watch, 81
New Economics Foundation, 58
New World Order, 32
NHS, 25; charges, 60; fundholding, 84; health centres, 83; hospitals, 24; Trusts, 84
Nitrogen tax, 106
Non-Governmental Organisations (NGOs), 93, 94, 95; see also Pressure Groups
Non-violent direct action, 2
Northern Ireland, 92
Nuclear companies, 32; disarmament, see Disarmament; energy, 33, 97; fallout, 43; power, 1, 31, 33, 34, 77, 89, 91, 94, 95, 102, 109; testing, 36; threat, 31; umbrella, 91; waste, 32, 43, 55; weapons, 34, 91
Nursing homes, 25, 83
Nutrition, 8, 73, 76, 77, 84, 93

Offenders, see Crime
OFGAS, 46
Oil, see Energy
Open University, see Education
Opinion polls, 43, 88
Opposition, 4, 5, 8, 12, 90, 111
Organic farming and food, see Agriculture
Organisation for African Unity (OAU), 78
Organisation for Security and Co-operation in Europe(OSCE), 90
Overseas Development Council (ODC), 38
Overseas Development Ministry, 94
Ozone, see Pollution
Ozone layer, 43

Packaging, 43, 101
Pakistan, 37, 38
Patents, 104, see also GATT
Peace, 71, 86, 87, 88, 89, 90, 95, 113; Dividend, 32, 90; movement, 31
Peace studies, see Education

Index

Peace-keeping force, see UN
Pension(s), 13, 83; funds, 17, 70, 106; State earnings-related pension(SERPS), 17
Pergau Dam, 35, 90,
Pesticide use, 77, 97
Photovoltaic cells, see Renewable Energy
Physical Impairment, 21
Plaid Cymru, 66
Planning, 49, 54, 79, 86, 104
Plutonium, 34
Pollution, 6, 16, 43, 46, 47, 49, 58, 59, 65, 84, 107, 108; acid rain, 46, 55, 101; marine, 55; ozone, 58, 101; toxic waste, 31, 107; ultra-violet radiation, 43, 44
Population, 1, 36, 37, 39, 45, 49, 57, 94
Porritt, Jonathon, 47
Poverty, 3, 8, 38, 48, 57, 59, 61
Poverty trap, 10
PowerGen, 109
Precautionary principle, 104
Pressure groups, 1, 41, 47, see also NGOs
Prevention, 83
Prison, 20, 29, 82, population, 20
Private sector, 16, 17, funding, 9, 16
Privatisation, 26
Proportional representation, see Electoral systems
Prostitution, 33
Protection, 6, 18, 20, 35, 58, 64, 70, 88, 94, 98; environmental, see Environmental
Public transport, see Transport

Quality of life, 12, 36, 58, 59
Quango, 26, 27, 79

Race relations, 78
Racism, 21; racial harassment, 86
Radioactive waste, see Nuclear waste
Re-use, 65, 100, 101
RECARM programme, see Disarmament
Recession, 11
Recycling, 65, 66, 100, 101
Referendum, 75, 77, 96, 97
Refugees, 8, 78, 95
Regional, see Government
Religion, 43, 56, 74
Renewable energy, 67, 102, 108, 109; geothermal power, 109; hydro-electric, 109; photovoltaic cell, 109; research & development, 102; solar, 67, 85, 109; tidal barrage; 109, wave power, 67, 109; wind power, 109
Repair, 62, 65, 72, 81
Repossessions, 11
Resource tax, see Taxation
Rights, 4, 17, 18, 29, 33, 37, 77, 78; animal, 36, 41, 75; asylum, 77; Bill of, 29, 76, 77, of the Child, 71; citizen's, 25; consumer, 25; employment, 13, 65; environmental, 76; human, 33, 35, 70, 75, 87, 92, 95, 96; intellectual property, 18; property, 32, to silence, 29; 'survival', 77; voting, 36, 77; to work, 78
Rio Summit, see UNCED
Road building, 80
RSPCA, 41
Rural development, 105
Russia, 33, 34, 89, 92
Rwanda, 95

Salination, 45, 48, 106
Scotland, 75
Second chamber, 76
Self-sufficiency, 94
Sellafield, 45
Sewage, 45, 48, 67, 107
Sex education, see Education
Sexism, 21
Sexual orientation, 77
Shareholder, 16, 70
Shopping, 24
Single European Act, see European Union
Single European Currency, see European Union
Single market, see European Union
Site of Special Scientific Interest (SSSI), 104
Sizewell, 45
Skills, 9, 21-23, 27, 40, 59, 62, 66, 72, 73, 82, 92; exchanges, 71; shortages, 14
Sleaze, 3
Small business, see Business
Social dumping, 18
Social housing, see Housing
Social justice, 4
Social Security, Department of (DSS), 9, 62
Social services, 75
Socialist, 4
Soil erosion, 48, 49, 51, 106; in Himalayas, 36
South America, 28, 40, 47
Soviet Union, 32, 33, 34, 87, 90
Speculation, 17
Speed limits, see Transport
Spiritual development, 74
Stakeholder, 84, 86
Steiner schools, 74
Stockholm Institute, 58
Stockmarket, 16
Strategic Arms Reduction Treaties (START), see Disarmament
Students, see Education
Subsidiarity, 6, 55, 57, 75

Sustainable, 16, 48, 53, 63, 65, 69, 70, 72; agriculture, 51, 104; development, 40, 53, 54, 55, 90, 93, 94; 'Ecological Sustainable Union', 97; economy, see Economy; growth, see Growth; society, 1, 14, 54, 76, 85, 113; trade, 108

Tax/Taxation, 13, 15, 61, 76; breaks, 61; carbon, 88, 102; currency, 69; eco-taxes, 99; employment, 102; energy, 97, 101, 102, 106; environmental, 60; incentives, 101; income, 60, 64; landfill, see Landfill; Land Value, see Land; low, 111; nitrogen, 106; packaging, 101; resource, 60, 64, 66, 100; system, 60, 66; VAT, 69; work, 60

Technology, 6, 12, 14, 67, 88, 91, 95, 99, 109; transfers, 88

Television, 13, 20, 31

Terrorist, 87

Thatcher, Mrs, 14, 27

Third World, see Developing Countries

Tobacco, 84

Topsoil, 49, 50

Torture, 5, 90

Tourism, agri-tourism, 106, eco-tourism, 103

Toxic waste, see Pollution

Trade, see Fair/ Free

Trade-related Intellectual Property Rights, see GATT

Trades Unions, 15, 18, 64, 65, 77, 90

Traffic calming, 81

Training, 2, 10, 13, 14, 22, 39, 63, 65, 72, 74, 78, 79

Transnational companies(TNCs), see Companies

Transport, 9, 15, 41, 46, 51, 75, 80, 84, 86, 102, 108, 111; buses, 80; cars, 19, 20, 80, 81, 108; cycling, 66, 80, 84; lorries, 80; public, 16, 20, 23; 24, 66, 67, 80; speed limits, 84; walking, 66, 80, 84; waterways, 80, 103, 107

Travellers, 2, 29, 77

Treasury, 9, 111

Trident, 91

Turkey, 16, 48

TVEI, see Education

Ukraine, 89

Ultra-violet radiation, see Pollution

Unemployment, automatic, 12

United Nations, UN Conference on Environment & Development(UNCED), 35; UN Convention on the Rights of the Child, 71; UN Development Programme (UNDP), 59, 90, 93, 103; UN Environment Programme, 103; UN Global Environment Facility, 69, 103; UN Security Council, 34, 95; UN Special Session on Disarmament, 73; UN World Summit on Social Development, 93,

United States of America, 8, 32, 42, 46, 49, 58, 84, 101, 103, 104,

United States of Europe, see Europe

Utilities, 16, 46, 48

VAT, see Taxation

Vigilante, 82

Voluntary; groups, 83, 109; organisations, 105; sector, 27, 79; work, 65, 78

Voting age, 76

Wage, minimum

Wales, 16, 27, 75

Walking, see Transport

Water, 1, 15, 16, 31-33, 36, 39, 42-45, 47-49, 51, 63, 67, 75-77, 85, 93, 99, 103, 106-109; table, 36, 45, 106

Waterways, see Transport

Wave power, see Renewable Energy

Weapons, 5, 32, 33, 34, 43, 81, 88, 89, 92; chemical, see Chemical; decommissioning, 89; nuclear, see Nuclear; procurement, 93

West Bank, 48

Western European Union, 91

Whole-cost accounting, 89

Wildlife, 46, 48, 105

Wind power, see Renewable Energy

Women, 5, 8-10, 13, 19, 20, 30, 37, 39, 40, 58, 75, 81, 100; women's movements, 54, 56

Work, ethic, 64; 'invisible', 58; 'good work', 64; guest, 38-39; low-paid, 10, 61; part-time, 13, 65

Work-fare, 64

World Bank, 93, 94, 103, 107

World Trade Organisation (WTO), 6, 93, 94

Xenophobia, 5

Young people, 2, 23, 39, 60, 61, 79, 84

Youth; facilities, 86; groups, 79; work, 80

Yugoslavia, ex, 35, 95

Zimbabwe, 109

Zoning, 79

Zoos, 103

Ethical Investment

A saver's guide
Peter Lang

The book for anyone with money to invest — whether a few hundred pounds, or many thousands — who tries to apply ethical standards to their everyday life. It is written in everyday language, free of the confusing jargon of the financial world.

Unlike the typical financial adviser, the author explains and describes all the ethical investment opportunities, including those that don't pay a commission to 'independent financial advisers' for recommending them. These include banks, building societies, and a number of funds and companies in the so-called 'social economy', as well as the commission-paying unit trusts, PEPs and pension funds. There is also a discussion of the choice of insurance companies.

£10 pbk   192pp   1 897766 20 3

Low Impact Development

Planning and people in a sustainable countryside
Simon Fairlie

This complete re-examination of Britain's planning system from the point of view of the planned, rather than of the planner, is an important contribution to the topical debate about the future and use of the countryside and what it means to achieve sustainability in the modern world.

Simon Fairlie argues that planners should look favourably on proposals for low impact, environmentally benign homes and workplaces in the open countryside. Criteria for planning approval at present favour the wealthy commuter and the large-scale farmer and discriminate heavily against (e.g.) smallholders, low-impact homes, and experimental forms of husbandry.

£10 pbk   176pp illustrated   1 897766 25 4

The Power in Our Hands

Neighbourhood based, world shaking
Tony Gibson

Telling the story of *ordinary* people doing *extraordinary* things, Tony Gibson uses worldwide examples to illustrate his argument that the potential for change lies in basic human assets: our creative instinct, our staying power, our urge to ask why? and how?, support from family and friends, and the time on our hands.

A power toolkit for neighbourhood development and political change, described by John Vidal as "the missing link in the debate about how we live, the secret formula that our leaders consistently misunderstand."

£10 pbk   320pp illustrated   1 897766 28 9

Also published by Jon Carpenter

Homelessness: What *Can* Be Done
An immediate programme of self-help and mutual aid
Ron Bailey

A practical programme for the provision of homes for the homeless, from one of the country's most experienced campaigners for the underprivileged.

Ron Bailey argues that central and local government cannot solve the problem of homelessness. Instead, he claims, the energy and enthusiasm of the homeless themselves can be harnessed in a massive self-build programme. The role of local government is to facilitate this effort. The book shows exactly how and where these homes can be built, and where such large sums of money will come from.

£7.99 pbk   112pp   1 897766 09 2

Towards a Sustainable Economy
The need for fundamental change
Ted Trainer

A lucid and hard-hitting analysis of the truth about our economic system that explains precisely why a few people are getting richer, most people are getting poorer, and why — if we don't change our ways — we're all heading for global catastrophe. Mass poverty and hunger, unemployment, under-development, waste, armed conflict, resource scarcity and environmental destruction are all caused by the disastrous flaws in our economy. Trainer shows how economic growth is seriously mistaken because it ignores finite resource and ecological limits, thereby promoting violence and injustice as well as ecological calamity.

£10.99 pbk   192pp   1 897766 14 9

Reinventing the Economy
The Third Way
David Simmons

How Margaret Thatcher and, after her, John Major have traded on popular disillusionment with socialism and a deliberate falsification of what Keynes was saying to discredit the left-of-centre consensus in Britain today and impose an unjust economic and social system that has but one objective: to re-allocate wealth from the lower and middle income groups to the already wealthy, by making ninety per cent of people worse off. This has been done at the expense of the job security and welfare of the overwhelming majority.

£11.99 pbk   288pp   1 897766 17 3

SEND FOR A FREE CATALOGUE OF OVER 100 BOOKS TO
JON CARPENTER PUBLISHING, THE SPENDLOVE CENTRE, CHARLBURY OX7 3PQ